Introductory Agroforestry

Theory and Practices

Introductory Agroforestry
Theory and Practices

As per 6th Dean Committee Recommendation of the ICAR

– Authors –

Prof. A. Zaman

Dean, School of Agricultural Sciences
Sister Nivedita University,
Kolkata, West Bengal

Dr. Md Hedayetullah

Assistant Professor, Agronomy,
BCKV, West Bengal

2026

Daya Publishing House®
A Division of
Astral International Pvt. Ltd.
New Delhi – 110 002

ISBN 978-93-5919-666-4

Published by : **Daya Publishing House**®
A Division of
Astral International Pvt. Ltd.
– ISO 9001:2015 Certified Company –
4736/23, Ground Floor, Ansari Road
Darya Ganj, New Delhi-110 002
Ph. 011-43549197, 8130496929
e-mail: info@astralint.com
Website: www.astralint.com

Printed at : **Replika Press Pvt Ltd**

Preface

In an era defined by unprecedented environmental challenges and the pressing need for sustainable food production, agroforestry emerges as a profoundly relevant and transformative discipline. It offers a powerful blend of ecological resilience, economic viability, and social equity by integrating trees with crops and/or livestock systems on the same land unit. This harmonious co-existence not only enhances productivity but also plays a pivotal role in biodiversity conservation, soil health improvement, water resource management, and climate change mitigation.

This textbook, "Introductory Agroforestry," is meticulously crafted to serve as a comprehensive foundational resource for students, practitioners, and enthusiasts seeking to understand the multifaceted world of agroforestry. Drawing directly from a well-structured syllabus, the book aims to bridge the gap between theoretical understanding and practical application, providing a holistic perspective on this dynamic field.

The theoretical chapters systematically delve into the core concepts, beginning with the fundamental definition and scope of agroforestry systems, exploring their diverse typologies, and highlighting the immense potential these systems hold, particularly within the varied agro-climatic zones of India. Readers will gain in-depth knowledge of Multipurpose Tree Species (MPTS)—their selection, vital role, and the intricate ecological aspects of tree-crop interactions, including competition and nutrient recycling. The book underscores the significance of traditional agroforestry as a viable choice for conserving India's rich agro-biodiversity and details the effective management strategies for agroforestry systems. Crucial environmental benefits, such as soil and water conservation, windbreaks, and shelterbelts, are thoroughly examined, alongside the critical socio-economic dimensions and the methodologies for designing and diagnosing agroforestry systems. A substantial section is dedicated to Silviculture, covering tree propagation techniques, nursery management, cultural practices, and the silvics of important tree species, including their specific requirements and utility. Furthermore, the book provides insights into practical agroforestry models developed by leading Indian research institutions like ICAR-IGFRI and the Indian Council of Forestry Research and Education (ICFRE).

Complementing the theoretical exposition, the practical modules are designed to cultivate essential hands-on skills. Students will learn to identify tree species, measure tree growth, understand environmental parameters affecting agroforestry systems, and master various plant propagation methods, including pre-sowing seed treatments, nursery bed preparation, and advanced techniques like pruning, coppicing, and pollarding. The curriculum includes practical training in afforestation methods, planting patterns, natural and artificial regeneration, and systematic procedures for the design, diagnostic survey, and evaluation of agroforestry systems across different agro-climatic zones. To provide real-world exposure, the book encourages both physical visits to prevailing agroforestry systems and related institutions within the state, as well as virtual tours of innovative models developed by premier research bodies.

Our aspiration is that this book will not only impart knowledge but also inspire a new generation of agroforesters and land managers to champion sustainable land use practices. By fostering a deep appreciation for the ecological, economic, and social benefits of integrating trees into agricultural landscapes, "Introductory Agroforestry" seeks to empower readers to contribute meaningfully to a more resilient and sustainable future.

Prof A. Zaman

Dr. Md. Hedayetullah

Contents

Preface *v*

Syllabuses as per 6th Deans' Committee Recommendation of the ICAR *xi*

INTRODUCTORY AGROFORESTRY (THEORY)

1. **Agroforestry: Definition and Scope of Agroforestry System; Distinctions between Agroforestry and Social Forestry** **3**
2. **Type of Agroforestry System, Potential of Agroforestry in India and Prevailing Agroforestry System in India** **11**
3. **MPTS: Definition, Role of MPTS in Agroforestry System, its Selection for different Agroforestry System, MPTS of India** **23**
4. **Ecological Aspects of Agroforestry System, Tree-Crop Interaction: Competition, Nutrient Recycling** **45**
5. **Traditional Agroforestry as a Viable Choice to Conserve Agro-biodiversity of India** **55**
6. **Management of Agroforestry System** **67**
7. **Role of Agroforestry in Soil and Water Conservation** **73**
8. **Windbreak and Shelterbelt–Definition, Objectives** **79**
9. **Socio-economic Aspects of Agroforestry System** **85**

10. Design and Diagnostic Study of Agroforestry System 91

11. Silviculture: Definition and Scope 107

12. Propagation of Tree Species, Management of Tree Species 113

13. Regeneration by Seed, Coppice, Root Suckers, Transplanting, Stump, Branch Cutting, Rhizomes 125

14. Nursery Bed Preparation and Management 137

15. Cultural Practices for Bare Root and Seedling, Field Handling of Nursery Stock 145

16. Silviculture of Important Tree Species, Choice of Species: Site Factors, Root, Crown and Bole Characteristics, Phenology, Nutritional and Water Requirement, Ground Operation, Tending, Harvesting Utility 151

17. Horticulture and Forage Crops-Based Agroforestry Models Developed by ICAR-IGFRI 157

18. Agroforestry Models Developed by Indian Council of Forestry Research and Education 163

INTRODUCTORY AGROFORESTRY (PRACTICAL)

1. Identification of Tree Species in Agroforestry 171

2. Study of Tree Growth Measurement 177

3. Study of Environmental Parameters Affecting Agroforestry System 181

4. Plant Propagation Methods 187

5. Pre-sowing Seed Treatment 193

6. Preparation of Nursery Bed Exercise 201

7. Practicing Propagation Techniques for Trees 207

8. Afforestation Method, Practical Training, Pruning, Coppicing and Pollarding 215

9. Planting Pattern and Designs for Plantation 227

10. Natural and Artificial Regeneration 237

11. Design and Diagnostic Survey of Agroforestry System 243

12. Evaluation of Agroforestry System in different Agroclimatic Zones **247**

13. Exposure Visit to Prevailing Agroforestry Systems of the State and Related Important Institutions **255**

14. Virtual Visit of Agroforestry Models Developed by ICAR-IGFRI, ICFRE **263**

Glossary **269**

References **273**

Index **275**

Syllabuses as per 6th Deans' Committee Recommendation of the ICAR

Introductory Agroforestry 2 (1+1)

Theory

Agroforestry: Definition and scope of Agroforestry system, Type of Agroforestry system, potential of Agroforestry in India, Prevailing agroforestry system in India; MPTS- definition, role of MPTS in agroforestry system, its selection for different agroforestry system, MPTS of India, Ecological aspects of Agroforestry system, tree -crop interaction – competition, nutrient recycling; Traditional Agroforestry as a viable choice to conserve Agro biodiversity of India. Management of Agroforestry system; Role of agroforestry in soil and water conservation; windbreak; Shelterbelt–definition, objectives.; Socio-economic aspects of Agroforestry system; Design and Diagnostic study of agroforestry system; Silviculture: Definition and scope, Propagation of tree species, Regeneration by seed, coppice, root suckers, Transplanting, stump, branch cutting, rhizomes; Nursery bed preparation and management; Cultural practices for bare root and seedling, field handling of nursery stock; Management of tree species; Silviculture of important tree species, choice of species- site factors, root, crown and bole characteristics, phenology, nutritional and water requirement, ground operation, tending, harvesting utility *etc.* Horticulture and forage-crops-based agroforestry models developed by ICAR-IGFRI; Agroforestry models developed by Indian council of Forestry Research and Education.

Practical

Identification of tree species in agroforestry, Study of tree growth measurement, Study of environmental parameters affecting AF System, Plant propagation methods, Pre-sowing seedtreatment, Preparation of nursery bed exercise, practicing propagation techniques for trees, Afforestation method, practical training, pruning, coppicing, pollarding *etc.* Planting pattern and designs for plantation, natural and artificial regeneration, Design and diagnostic survey of agroforestry system, Evaluation of agroforestry system in different agro climatic zones, Exposure Visit to prevailing agroforestry systems of the state and related important institutions, Virtual visit of agroforestry models developed by ICAR-IGFRI, ICFRE.

The Authors

Professor (Dr.) A. Zaman, Ph D in Agronomy, the esteemed author of *Deeksharambh* (*Induction-cum-Foundation Course in Agriculture*), is an internationally recognized scientist with 40 years of profound experience in agricultural water management. His ground-breaking and original research in fields such as irrigation water management, rainfed agriculture, cropping systems, watershed management, dry land agriculture, and biodiversity conservation has garnered global appreciation and established him as an outstanding teacher within farm universities worldwide. Dr. Zaman is credited with developing a wealth of original concepts, strategies, techniques, and tools that have profoundly contributed to agricultural science and benefited global society.

Born May 27, 1954, is a distinguished Agronomist specializing in Agricultural Water Management for enhancing crop and water productivity. His extensive career spans influential leadership roles at Bidhan Chandra Krishi Viswavidyalaya (BCKV) and Uttar Banga Krishi Viswavidyalaya (UBKV) in West Bengal, India. He has notably served as Dean, Faculty of Agriculture; Director of Research; Head of the Department of Agronomy; and Chief Scientist in AICRP on Water Management (ICAR). With significant research experience in water management, a background in post-graduate teaching, and considerable administrative acumen,

Since May 2020, Prof. Zaman has served as the Dean of the School of Agricultural Sciences at Sister Nivedita University, Kolkata. His extensive professional journey includes pivotal roles as Founder In-charge of Agriculture at The Neotia University (June 2018 to 2020), Emeritus Professor in Agriculture at M.S. Swaminathan School of Agriculture, Centurion University of Technology and Management (September 2016 – May 2018), and Programme Coordinator for the World Bank-funded West Bengal Accelerated Development of Minor Irrigation Project (September 2014 – August 2016).

Prior to these roles, Dr. Zaman gained invaluable international experience as Head of the Irrigation Expatriate Team for the Government of Uganda (1994-1996) under an ITEC Programme. In this capacity, he was instrumental in designing, laying out, and installing farmer-managed small-scale irrigation systems for sustainable crop cultivation. His vast academic contributions include authorship of 10 textbooks in agriculture, several book chapters, over 150 peer-reviewed research papers, and 179 publications in national and international seminar and conference proceedings. Furthermore, he has significantly shaped future researchers, having guided 7 Ph.D. candidates and supervised over 50 postgraduate thesis works.

Dr. Md Hedayetullah, PhD, is a highly respected expert in agricultural science and the co-author of "Deeksharambh (Induction - cum-Foundation Course in Agriculture)." He is currently an Assistant Professor/Scientist and Officer-In-Charge of the ICAR's All India Coordinated Research Project on Chickpea at Bidhan Chandra Krishi Viswavidyalaya (BCKV), Kalyani, West Bengal.

Holding a PhD in Agronomy from BCKV and a recipient of the Maulana Azad National Fellowship, Dr. Hedayetullah brings a wealth of practical and academic experience to this foundation course. His background includes roles as Agronomist at NABARD and Assistant Professor at the M.S. Swaminathan Institute of Agriculture Science and the College of Agriculture, Tripura, giving him a comprehensive understanding of agricultural challenges and solutions.

A prolific contributor to agricultural literature, Dr. Hedayetullah has authored or co-authored 6 books, 30 research papers, and numerous book chapters, covering a wide array of topics crucial for aspiring agriculturists. His involvement as Principal Investigator in various research projects and his extensive outreach through live TV talks and farmer training programs underscore his commitment to advancing agricultural knowledge and practice. His expertise makes him an ideal guide for students embarking on their agricultural journey.

Introductory Agroforestry (Theory)

Chapter 1

Agroforestry: Definition and Scope of Agroforestry System; Distinctions between Agroforestry and Social Forestry

Concept of Agroforestry

Agroforestry, the integration of trees with crops and/or animals, is an ancient land management practice deeply rooted in many cultures. Scientific understanding of its benefits is more recent, but farmers, particularly in warmer climates, have long recognized the advantages of cultivating food crops, trees, and livestock together for diverse yields. In India, trees and forests hold a particularly significant place, interwoven with cultural identity. Indeed, much of Indian philosophy originated in forest-dwelling communities where rishis lived in harmony with nature. Ancient literature emphasizes the importance of tree planting, with individuals often incorporating trees into their agricultural practices. The "Krishishukti" by Maharishi Kashyap, for example, identifies specific land categories, including wet and dry areas around homes, wells, and tanks, as suitable for tree planting.

Historically, foresters and agriculturists often focused on monoculture production, working within strict disciplinary boundaries and overlooking the potential of these integrated systems. However, deforestation and resource depletion have led to shortages of essential forest products like firewood, fodder, and timber, driving up prices and creating challenges for forest-based industries. This scarcity has spurred renewed interest in Agroforestry, with many farmers now incorporating trees into their farmlands to supplement agricultural crops.

Agroforestry, therefore, is a land-use system characterized by:

- ✰ The combination of trees with crops and/or animals on the same land unit.
- ✰ The integration of multiple outputs with the protection of the production base.
- ✰ The emphasis on the use of diverse indigenous trees and shrubs.
- ✰ Its suitability for low-input conditions and fragile environments.
- ✰ The interplay of sociocultural values.
- ✰ Greater structural and functional complexity compared to monoculture.

While the term "Agroforestry " is relatively new, the practices it encompasses are old. Many definitions exist, but it is now widely accepted as a sustainable land management system. For example:

- ✰ **Bene *et al.* (1977):** "A sustainable management system for land that increases overall production, combines agriculture crops, forest plants and tree crop and/or animals simultaneously or sequentially and applies management practices that are compatible with the cultural patterns of a local population."
- ✰ **King and Chandler (1978):** "A sustainable land management system which increases the overall yield of the land, combines the production of crops (including tree crops) and forest plants and/or animals simultaneously or sequentially, on the same unit of land and applies management practices that are compatible with the cultural practices of the local population."
- ✰ **Nair (1979):** "A land-use system that integrates trees, crops and animals in a way that is scientifically sound, ecologically desirable, practically feasible and socially acceptable to the farmers."
- ✰ **Lundgren and Raintree (1982):** "A collective name for land use systems and technologies, where woody perennials (trees, shrubs, palm bamboos, *etc.*) are deliberately used in the same piece of land management units as agriculture crops and/or animals in some form of spatial arrangement or temporal sequence. In Agroforestry systems, there are both ecological and economical interactions between the different components."

Based on these definitions, key characteristics of Agroforestry systems include:

- ✰ Involvement of two or more plant (or plant and animal) species, at least one of which is woody.
- ✰ Production of multiple outputs.
- ✰ A cycle longer than one year.

- ✰ The presence of both positive and negative interactions between components (trees, crops, animals).
- ✰ Ecological (structurally and functionally) and economic complexity exceeding that of monocropping systems.

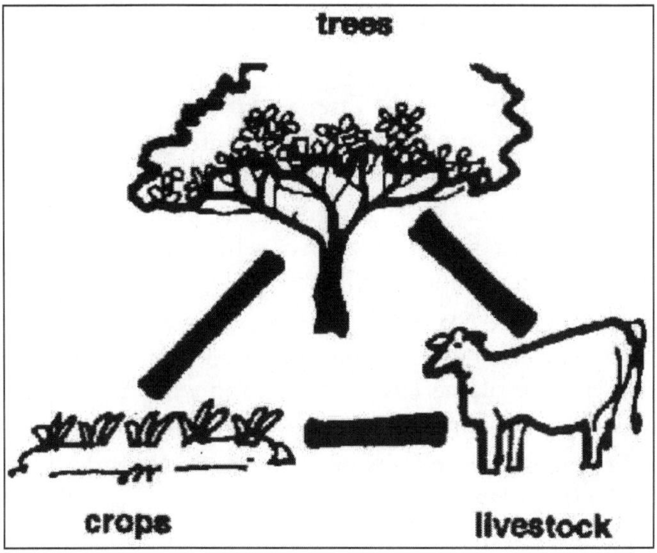

Figure 1.1: Agroforestry Components or Basic Components of Agroforestry.

Agroforestry land management aims to simultaneously conserve and improve the site while optimizing the combined production of trees, agricultural crops, and livestock.

Agroforestry systems, in theory, possess three key attributes:

- ✰ **Productivity:** Agroforestry seeks to maintain or increase the production of preferred commodities and overall land productivity. This can be achieved through increased tree product output, improved crop yields, reduced input requirements for cropping systems, and enhanced labor efficiency.
- ✰ **Sustainability:** By leveraging the beneficial effects of woody perennials on soils, Agroforestry can conserve resources and maintain long-term soil fertility.
- ✰ **Adoptability:** Agroforestry practices should be readily accepted by local farming communities. New or improved Agroforestry technologies must align with existing farming practices to ensure successful implementation.

Agroforestry holds significant potential to address the demands of a growing population by:

- **Meeting Food, Fuelwood, and Fodder Needs:** Enhanced food production through nitrogen fixation, improved nutrient access, and mycorrhizal associations; direct food sources from trees (fruits, nuts); and fodder for livestock.
- **Conserving Water:** Improved soil moisture retention, stream flow regulation, reduced flood hazards, and drainage of waterlogged or saline soils.
- **Providing Fuelwood and Energy:** Fuelwood for direct combustion, pyrolytic conversion products (charcoal, oil, gas), ethanol from fruit fermentation, and combustible oils/resins.
- **Providing Shelter:** Building materials, shade for people/livestock/crops, windbreaks for protection of settlements/crops/roadways, and fencing (live fences and fence posts).
- **Supplying Raw Materials for Industries:** Raw materials for pulp and paper, tannins, essential oils, medicinal ingredients, wood for agricultural implements and crafts, and weaving fibers.
- **Generating Cash Benefits:** Direct income from tree product sales, indirect income from increased productivity, and higher land value compared to monoculture.
- **Increasing Yield and Maximizing Production:** Year-round solar radiation utilization, nitrogen fixation by leguminous trees, and leaves as green manure for improved soil productivity.
- **Diversifying Products:** A variety of food materials from trees, shrubs, herbs, and climbers for sustenance; production of fuel, fodder, fruits, fiber, and timber to meet diverse needs.
- **Utilizing Wasteland and Degraded Land:** Restoring degraded lands through vegetation cover, ecological restoration, and soil improvement for subsequent agricultural production.
- **Creating Employment Opportunities:** Employment generation through plantation activities, seed collection, nursery raising, and wood-based industries.
- **Providing Carbon Sequestration Services:** Responding to climate change through carbon sequestration in plant biomass and soils.
- **Reducing Deforestation Rates:** Reducing pressure on natural forests by providing readily available fuelwood and decreasing the time spent foraging for fuel.
- **Improving Soil Health:** Improving soil's physico-chemical properties and fostering favorable environments for soil microbes and fauna.
- **Creating Habitats for Wild Species:** Enhancing connectivity and landscape heterogeneity in multi-functional conservation landscapes.

Social forestry is a forestry practice that takes place outside of traditional forest areas, aimed at benefiting both rural and urban communities. The term was first coined by J.C. Westoby and gained recognition in the interim report of the National Commission on Agriculture (NCA) in 1976 as a crucial component of forestry for meeting rural needs. The NCA established the objectives of social forestry to fulfill the basic and economic needs of the community, with a scope that included farm forestry, community woodlots, and reforestation of degraded lands.

Social forestry encompasses several areas, including:

1. Farm Forestry: This involves forestry practices on farms, such as planting rows of trees on bunds or boundaries of fields, individual trees in private agricultural land, and creating wind breaks for protection.
2. Extension Forestry: This refers to forestry practices in areas devoid of tree growth, with the objective of increasing the area under tree cover. It includes mixed forestry, shelterbelts, wind breaks, linear strip plantations, and community woodlots.
   - ✰ Mixed forestry involves raising fodder grass with scattered fodder, fruit, and fuel-wood trees on suitable waste lands, panchayat land, and village commons land.
   - ✰ Shelterbelts are extensive belts of trees and/or shrubs maintained for the purpose of shelter from wind, sun, snow-drift, *etc.*, covering areas larger than a single farm and sometimes whole regions on a planned pattern.
   - ✰ Wind breaks are protective plantings around a garden, farm, or field to protect it against strong winds.
   - ✰ Linear Strip Plantation involves plantations of fast-growing species on linear strips of land on the sides of public roads, canals, and railway lines.
   - ✰ Community Woodlots: These consist of plantations of fuelwood species on community village lands for increasing a villager's access to fuel wood, fruits, and fodder.
3. Rehabilitation of Degraded Forests: This involves reforestation of degraded forests to meet fuel and timber requirements, organize fuelwood supplies at reasonable rates, provide employment, and rehabilitate degraded forests.
4. Recreation Forestry: This refers to the practice of forestry for raising avenue/flowering trees and shrubs mainly to serve as recreation forests for the urban and rural population, also known as aesthetic forestry.

Social Forestry

Distinction between Social Forestry and Agroforestry

Social Forestry	*Agroforestry*
1. Social forestry is a plantation made on lands outside conventional forest areas for the benefit of rural and urban communities, with objectives to supply fuel wood to divert cow dung from village hearths to village fields, small timber for housing and agricultural implements and fodder for cattle of the rural population, protection of agriculture by creation of diverse ecosystem and arresting wind and water erosion, provide raw material for village cottage industries and improve scenic value in rural and urban areas.	1. Agroforestry is a sustainable land management system that increases the overall production, combines agricultural crops, tree crops and forest plants and/ or animals simultaneously or sequentially, and applies management practices that are compatible with the cultural patterns of the local population.
2. It is thus the forestry of the people, by the people and for the people.	2. It is a system which is rather localized in its concept for managing the unit of land for maximizes production of agricultural crop and forest trees complimentary with each other.
3. Planting of trees on massive scale is done on vacant land community land, roadside railway track and even degraded reserve forest. Helps to eradicate poverty especially among land less and marginal rural people by providing them job potential.	3. Agroforestry is practiced mostly in farmers' field/ownland.
4. Mainly trees and shrubs are to be used to harvest multiple products.	4. It involves integration of two or more than two components ion the same unit of land.
5. Social forestry is primarily a government based programme that aims to increase the forest area by rehabilitating wastelands while producing biomass both for industrial and local uses.	5. Agroforestry involves the rural awakening towards self sufficiency by producing maximum biomass per unit area, fulfilling then needs of food, fodder and fuelwood *etc.*

QUESTION BANK

1. **What is the correct definition of Agroforestry?**
   A. A system involving only crop cultivation
   B. A system that integrates trees, crops, and/or livestock on the same land unit
   C. Plantation of forests in reserved forest areas
   D. Farming without the use of synthetic inputs

Answer: B. A system that integrates trees, crops, and/or livestock on the same land unit

2. **Which of the following is NOT a component of Agroforestry?**
   A. Agricultural crops
   B. Livestock
   C. Urban landscaping
   D. Trees or woody perennials

Answer: C. Urban landscaping

3. **The primary distinction between Agroforestry and Social Forestry is that Agroforestry is practiced mainly on:**
   A. Community lands
   B. Institutional lands
   C. Private farmlands
   D. Government forest lands

Answer: C. Private farmlands

4. **Social Forestry mainly aims to:**
   A. Maximize crop productivity
   B. Increase soil erosion
   C. Meet local community needs for fuelwood, fodder, and small timber
   D. Promote commercial timber production

Answer: C. Meet local community needs for fuelwood, fodder, and small timber

5. **Which of the following best describes the scope of Agroforestry?**
   A. Only timber production
   B. Replacing crops with trees
   C. Sustainable land use combining agriculture and forestry
   D. Growing monoculture plantations

Answer: C. Sustainable land use combining agriculture and forestry

6. One major goal of Agroforestry is to:

A. Eliminate the need for fertilizers

B. Convert all land into forest

C. Optimize productivity through biological interactions between components

D. Reduce landholding size

Answer: C. Optimize productivity through biological interactions between components

7. A key difference between Agroforestry and traditional agriculture is:

A. Agroforestry uses more machinery

B. Agroforestry includes woody perennials

C. Traditional agriculture is always more profitable

D. Agroforestry uses no soil

Answer: B. Agroforestry includes woody perennials

8. Which type of land is primarily targeted in Social Forestry programs?

A. Private cultivated land B. Forest reserve areas

C. Community and wastelands D. Commercial plantations

Answer: C. Community and wastelands

9. In terms of policy, Social Forestry programs are mainly initiated by:

A. Private farmers B. NGOs

C. Government and local bodies D. Export industries

Answer: C. Government and local bodies

10. Which of the following practices is most characteristic of Agroforestry systems?

A. Seasonal monocropping

B. Tree-crop-livestock interaction on the same land

C. Reforestation of degraded forest areas

D. Large-scale logging operations

Answer: B. Tree-crop-livestock interaction on the same land

Chapter 2

Type of Agroforestry System, Potential of Agroforestry in India and Prevailing Agroforestry System in India

Agroforestry in India: An Overview of Systems and Potential

Introduction

Agroforestry is the intentional integration of trees and/or shrubs into crop and/or animal farming practices, creating diverse and productive land-use systems. It is a traditional farming method practiced in many parts of the world, including India. Agroforestry systems provide a multitude of benefits, including increased agricultural productivity, improved soil health, enhanced biodiversity, and reduced vulnerability to climate change. This article explores the different types of agroforestry systems, the potential of agroforestry in India, and the prevailing agroforestry systems in the country.

Types of Agroforestry Systems

1. Silvopasture: This system involves the integration of trees, crops, and animals on the same piece of land. The trees provide shade, shelter, and food for the animals, while the animals help control weeds and provide manure for the trees and crops.
2. Silviculture: This system involves growing trees on agricultural lands for timber, fuelwood, or other wood products. The trees can be grown in between or above crops, providing additional income for farmers.

3. Alley Cropping: This system involves growing crops in between rows of trees or shrubs. The trees or shrubs provide shade, windbreaks, and soil conservation, while the crops provide food and income for the farmers.
4. Forest Farming: This system involves growing specialty crops, such as medicinal plants, spices, and fruits, under a forest canopy. The forest provides a natural habitat for the crops, reducing the need for chemical inputs.

Potential of Agroforestry in India

India has a vast potential for agroforestry, with its diverse agro-climatic zones, varied topography, and rich biodiversity. The country has more than 150 million hectares of agricultural land, of which only a small fraction is currently under agroforestry. However, the potential for agroforestry in India is enormous, as it can help address some of the country's pressing challenges, such as food security, climate change, and rural poverty.

Agroforestry can increase agricultural productivity, particularly in areas where crop yields are low due to poor soil health or water scarcity. The trees in agroforestry systems can improve soil fertility, reduce soil erosion, and enhance water availability, leading to higher crop yields. Moreover, agroforestry can provide additional income for farmers through the sale of timber, fuelwood, and other wood products.

Agroforestry can also help mitigate climate change by sequestering carbon dioxide from the atmosphere. The trees in agroforestry systems can absorb and store large amounts of carbon, reducing greenhouse gas emissions and helping to combat global warming.

Prevailing Agroforestry Systems in India

Agroforestry has been practiced in India for centuries, particularly in the tribal and rural areas of the country. The prevailing agroforestry systems in India include:

1. Taungya System: This system involves the cultivation of crops between newly planted trees in a forest plantation. The crops are grown for a few years, after which the trees are harvested for timber.
2. Homegardens: This system involves the cultivation of a variety of crops and trees around the homestead. Homegardens are common in the southern and western parts of India and provide food, medicine, and income for the household.
3. Agri-horticulture: This system involves the integration of fruit trees with crops. The trees provide food, shade, and income, while the crops provide food and income for the farmers.

4. Agri-silviculture: This system involves the integration of trees with crops. The trees provide timber, fuelwood, and other wood products, while the crops provide food and income for the farmers.

Agroforestry has the potential to transform India's agricultural landscape, providing multiple benefits to farmers, the environment, and society at large. The country's diverse agro-climatic zones and rich biodiversity provide an excellent opportunity for the widespread adoption of agroforestry practices. By promoting agroforestry, India can increase agricultural productivity, enhance rural livelihoods, mitigate climate change, and conserve biodiversity.

However, the adoption of agroforestry practices in India is not without challenges. There is a need for greater awareness and understanding of the benefits of agroforestry among farmers, policymakers, and the general public. Moreover, there is a need for supportive policies and regulations that encourage the adoption of agroforestry practices.

In conclusion, agroforestry has the potential to transform India's agricultural landscape, providing multiple benefits to farmers, the environment, and society at large. By promoting agroforestry, India can increase agricultural productivity, enhance rural livelihoods, mitigate climate change, and conserve biodiversity. It is time for policymakers, researchers, and farmers to come together and harness the potential of agroforestry in India.

Defining a Forest

A forest can be defined in multiple ways. From a management perspective, it's an area designated for timber and other forest product production, or maintained with trees and woody vegetation to provide indirect benefits such as climate regulation and environmental protection. Ecologically, a forest is a plant community primarily composed of trees and other woody plants, often characterized by a closed canopy. Legally, a forest is land formally designated as such under relevant forest laws.

Forest Stands

Within a forest, a stand refers to a group of trees occupying a specific area. These trees share relatively uniform characteristics in terms of species composition, age, arrangement, and overall condition, making the stand distinct from surrounding areas.

Forest Classification

Forests are classified according to several criteria, including: the method of regeneration (natural seeding or vegetative reproduction), age structure (even-aged or uneven-aged), species composition (pure or mixed stands), management objectives, ownership and legal status, and the characteristics of the existing trees (growing stock).

Classification by Regeneration Method

- **High Forests:** Regenerated primarily from seeds.
- **Coppice Forests:** Regenerated vegetatively through coppicing (shoots from cut stumps) or root suckers.

Classification by Age Structure

- **Even-Aged Forests:** Consist of trees of roughly the same age, with age differences generally less than 25 per cent of the rotation age (the planned period between harvests).
- **Uneven-Aged Forests:** Composed of trees with markedly different ages, exceeding 25 per cent of the rotation age.

Classification by Species Composition

- **Pure Forests:** Dominated by a single tree species, typically comprising at least 80 per cent of the trees.
- **Mixed Forests:** Contain two or more tree species intermingled within the same canopy.

Forests can be classified according to their management goals, ownership, and growing stock.

Management Objectives

- Productive Forests: Managed mainly for timber and other forest products, these are sometimes called national forests and are used to support defense, industry, communication, and other public needs.
- Protective Forests: Primarily managed to protect environmental resources such as regulating water flow, preventing soil erosion, and stabilizing shifting sands.
- Farm Forests: Trees planted on farms and adjacent areas, either individually or in groups, to provide fuel, fodder, and other benefits to agricultural operations.
- Fuel Forests: Planted on village wastelands to supply fuel, small timber, and fodder to communities far from government forests.
- Recreational Forests: Managed specifically to provide recreational opportunities for both urban and rural populations.

Ownership and Legal Status

- State Forests: Forests owned by the state government, classified as:
  - Reserved Forests: Designated under the Indian Forest Act of 1927 (or other relevant forest laws) with strict regulations.

- Protected Forests: Subject to a less restrictive level of protection under the Indian Forest Act of 1927.
- Village Forests: State forests assigned to a village community, as defined by the Indian Forest Act of 1927.
- Communal Forests: Owned and managed by a community (village, town, tribal authority, *etc.*), with produce shared among members.
- Panchayat Forests: Managed by a village panchayat, an elected body responsible for local administration.

Growing Stock

- Normal Forests: An idealized forest state for a specific site and management objective, characterized by a balanced growing stock, age class distribution, and growth rate, allowing for sustainable harvesting equal to the annual increment.
- Abnormal Forests: Forests that deviate from the normal state, exhibiting deficits or excesses in growing stock and/or an unbalanced distribution of age or size classes.

Indian Forest Status (2009)

- Forest cover in India was approximately 690,899 km2 (69.09 million ha), representing 21.02 per cent of the country's geographical area.
    - Very Dense Forest: 83,510 km2 (8.35 million ha) - Canopy density 70 per cent and above.
    - Moderately Dense Forest: 319,012 km2 (31.90 million ha) - Canopy density between 40 per cent and 70 per cent.
    - Open Forest: 288,377 km2 (28.84 million ha) - Canopy density between 10 per cent and 40 per cent.
    - Scrub: 41,525 km2 (4.15 million ha) - Degraded forest with canopy density less than 10 per cent.
    - Mangrove: 4,639 km2 (0.46 million ha).
- Excluding areas above the tree line (18.31 million ha), forest cover was 22.26 per cent.
- Hills and tribal districts, particularly in the North-Eastern states, showed significant increases in forest cover.
- Madhya Pradesh had the largest area under forest cover, followed by Arunachal Pradesh, Chattisgarh, Orissa, and Maharashtra.
- Mizoram had the highest proportion of its geographic area under forest cover, followed by Nagaland.
- North-East region accounted for 25.11 per cent of the nation's forest cover.

- ✰ Native forests in India were reportedly disappearing at a rate of up to 2.7 per cent per year.
- ✰ Non-forest areas are defined as any regions not classified as forest.

Forest Types of India

Champion and Seth (1967) categorized the forests of India into 5 major groups, which are further subdivided into 16 types.

Major Groups

1. Tropical Forests

This group includes seven subtypes:

- ✰ Wet Evergreen Forest
- ✰ Semi-Evergreen Forest
- ✰ Moist Deciduous Forest
- ✰ Littoral and Swamp Forest
- ✰ Dry Deciduous Forest
- ✰ Thorn Forest
- ✰ Dry Evergreen Forest

2. Montane Subtropical Forests

This group consists of three subtypes:

- ✰ Broad-Leaved Hill Forest
- ✰ Pine Forest
- ✰ Dry Evergreen Forest

3. Montane Temperate Forests

This group includes three subtypes:

- ✰ Montane Wet Temperate Forest
- ✰ Himalayan Temperate Forest
- ✰ Himalayan Dry Temperate Forest

4. Sub-Alpine Forests

This group has one subtype:

- ✰ Sub-Alpine Forest

5. Alpine Scrub

This group includes two subtypes:

- ☆ Moist Alpine Scrub
- ☆ Dry Alpine Scrub

Role of Forests in Farming Systems

Forests act as the world's natural air conditioner and the Earth's protective blanket, making the planet habitable. They are essential for environmental stability and economic benefits. Among all ecosystems, forests are the largest, most complex, and self-sustaining. Preserving forests is crucial for all communities, irrespective of their level of development, as they perform a wide range of functions. Key functions include:

1. Productive Functions of Forests

Forests are vital natural resources that supply valuable goods. Wood, as a primary forest product, is used for numerous purposes.

- ☆ Wood serves as a universal fuel. For centuries, it was humanity's main energy source before the rise of coal, oil, gas, and electricity. Today, over half of global wood consumption is still for fuel, and in India, it remains a key domestic energy source. Of the total demand for 201 million tonnes of fuelwood, about 103 million tonnes (51 per cent) come from forest areas and plantations, with the rest sourced from farm forestry and other lands.
- ☆ Forests provide raw materials for various industries, such as paper and pulp, plywood and boards, sawmills, furniture, packaging, matches, and toys.
- ☆ Out of the 64 million cubic meters of timber demand, approximately 31 million cubic meters are supplied by farm forestry and other woodlands, 12 million cubic meters by forests, and the remaining 21 million cubic meters from plantations and natural forests—mostly as smaller timber for household needs.
- ☆ Forests also yield numerous non-wood products, often referred to as minor forest products due to their smaller harvest volumes, though they are highly significant. Key examples include:
  - ❒ Fibers and flosses
  - ❒ Grasses and bamboos
  - ❒ Essential oils
  - ❒ Oilseeds
  - ❒ Tans and dyes
  - ❒ Gums and resins

- Drugs, spices, and insecticides
- Tendu and other leaves
- Edible products
- Lac and other materials
- Fodder and grazing resources

2. Protective and Ameliorative Functions of Forests

Forests contribute to environmental protection in several ways:

- They help maintain the balance of carbon dioxide (CO2) in the atmosphere. Without adequate forest cover, excess CO2 could lead to global warming, climate disruptions, melting polar ice, and rising sea levels. Currently, atmospheric CO2 levels have reached 0.042 per cent, exceeding the normal 0.030 per cent, which could cause severe global issues if unchecked.
- Forests enhance local rainfall by 5 to 10 per cent through their influence on orographic and microclimate conditions, promoting cloud formation.
- They lower temperatures and increase humidity; forest areas are typically 3°C to 8°C cooler than nearby open spaces, reducing evaporation and improving comfort. This cooling effect extends beyond forest boundaries.
- Forests preserve soil fertility by adding organic matter and recycling nutrients. Tree leaves can be used as manure, and the availability of firewood from forests allows animal dung to be repurposed as fertilizer.
- Tree canopies soften the impact of rain, reducing splash erosion, while improving soil infiltration and water retention to minimize surface runoff and soil loss.
- Forests prevent flooding by conserving soil and water, extending the water cycle and ensuring a steady stream supply. They promote slower subsurface runoff, which avoids erosion.
- Forests and trees reduce wind speed, curbing wind erosion, stabilizing sand dunes, and preventing desertification.
- As shelterbelts and windbreaks, they protect agricultural crops, especially in arid and semi-arid regions, leading to higher farm yields.

3. Recreational and Educational Functions of Forests

Forests offer various non-material benefits:

- They provide recreational opportunities, attracting people with their diverse trees, shrubs, animals, and birds. Popular sites include national parks and wildlife sanctuaries, which draw large crowds annually.

- ✰ Forests serve as living laboratories for students from colleges and universities, offering spaces for ecological research and experiments.
- ✰ Forests have a natural healing effect, promoting mental and physical well-being through their serene environments.

Developmental Functions of Forests

Forests play a vital role in socioeconomic development through a variety of functions:

1. **Employment Generation:** Forestry activities are generally labor-intensive, creating significant employment opportunities in both primary (*e.g.*, planting, harvesting) and secondary (*e.g.*, processing) sectors.
2. **Improved Socioeconomic Conditions for Tribals:** Forests provide tribals with opportunities to improve their livelihoods through the collection, processing, and marketing of forest products, supported by gainful employment. Forestry actively contributes to poverty alleviation programs in these communities.
3. **Government Revenue:** Forests generate substantial revenue for governments, which can then be invested in various developmental projects and initiatives.
4. **Soil Rejuvenation:** Forests promote biological soil health. Tree root systems aerate the soil and the addition of organic matter (litter/humus) improves soil fertility, creating a hospitable environment for beneficial microorganisms and fauna.
5. **Subsistence Products:** Forests provide essential subsistence products. They contribute significantly to livestock needs, supplying a considerable portion of the country's fodder requirements, with substantial amounts of dry and green fodder extracted annually.
6. **Enhanced Agricultural Productivity:** In semi-arid regions, trees enhance soil productivity and land sustainability by recycling nutrients and providing mulch and shade for crops, thus complementing agricultural production.
7. **Soil Enrichment and Erosion Control:** Maintaining trees on farms widely benefits soil health and reduces erosion.
8. **Supplementary Income:** Planting trees on farm boundaries or intercropping them with field crops can generate supplementary income for farmers without significantly impacting the yield of their primary crops.
9. **Increased Overall Income:** Homestead plantations increase the overall income generated from land.

10. **Land Use Shift for Economic Stability:** In situations where agricultural income is unreliable or insufficient, and there are limited off-farm employment opportunities, small farmers may shift their land use to tree plantations. Trees require less labor, allowing farmers to focus on wage labor to meet immediate consumption needs.
11. **Profit Maximization:** Trees can be substituted for agricultural crops to increase overall profit from land.

QUESTION BANK

1. **Which of the following is a major type of Agroforestry system?**

   A. Hydrophonics B. Agrisilviculture

   C. Sericulture D. Aquaculture

Answer: B. Agrisilviculture

2. **The combination of trees, crops, and livestock in a single land-use system is known as:**

   A. Silvopastoral system B. Agrihorticultural system

   C. Agrisilvipastoral system D. Hortiagriculture

Answer: C. Agrisilvipastoral system

3. **In India, which Agroforestry system is most prevalent in arid and semi-arid regions?**

   A. Agrisilviculture B. Silvopastoral

   C. Agrihorticulture D. Taungya system

Answer: A. Agrisilviculture

4. **Which tree species is commonly used in agroforestry systems of northwestern India for its drought resistance and soil-enriching properties?**

   A. Pinus roxburghii B. Prosopis juliflora

   C. Acacia nilotica D. Azadirachta indica

Answer: C. Acacia nilotica

5. **What is the primary objective of the Silvopastoral system?**

   A. Timber production

   B. Soil erosion control

   C. Integration of trees with pasture for livestock

   D. Aquatic biodiversity conservation

Answer: C. Integration of trees with pasture for livestock

6. **Which agroforestry system is practiced on forest lands involving cultivation along with forest trees, especially under government supervision?**

   A. Social Forestry B. Taungya System

   C. Silviculture D. Urban Forestry

Answer: B. Taungya System

7. **One of the key potentials of Agroforestry in India is:**
   A. Increasing urban migration
   B. Decreasing biodiversity
   C. Enhancing carbon sequestration and land productivity
   D. Promoting monoculture

Answer: C. Enhancing carbon sequestration and land productivity

8. **What is the estimated potential area for Agroforestry in India, according to government reports?**
   A. 13 million ha
   B. 26 million ha
   C. 80 million ha
   D. 120 million ha

Answer: B. 26 million ha

9. **In humid and sub-humid regions of India, which agroforestry system is common due to high rainfall and fertile soils?**
   A. Silvopastoral system
   B. Agrihorticultural system
   C. Agrisilviculture system
   D. Alley cropping

Answer: B. Agrihorticultural system

10. **Which is NOT a recognized advantage of Agroforestry in the Indian context?**
   A. Soil fertility improvement
   B. Employment generation
   C. Rapid urbanization
   D. Climate resilience

Answer: C. Rapid urbanization

Chapter 3

MPTS: Definition, Role of MPTS in Agroforestry System, its Selection for different Agroforestry System, MPTS of India

Introduction

Agroforestry is a land use management approach that combines trees and/or shrubs with crops and/or livestock on the same piece of land. This integration leads to numerous benefits, including increased productivity, improved soil health, diversified income sources, and enhanced ecosystem services. A crucial component of agroforestry systems is the use of Multi-Purpose Tree Species (MPTS). This article defines MPTS, discusses their role in agroforestry systems, outlines factors to consider for their selection, and provides examples of MPTS in India.

Definition of Multi-Purpose Tree Species (MPTS)

MPTS are tree or shrub species that provide multiple benefits and services, including wood products, non-wood products, ecological services, and socio-economic values. These benefits can be categorized into timber, fuelwood, fodder, food, medicine, shade, soil conservation, and wildlife habitat, among others. The versatility of MPTS makes them ideal for agroforestry systems, where land use efficiency and sustainability are critical objectives.

Role of MPTS in Agroforestry Systems

MPTS play a pivotal role in agroforestry systems by:

1. Enhancing productivity: MPTS can increase crop yields by providing shade, reducing wind erosion, and improving soil structure and fertility.
2. Diversifying income sources: MPTS can generate additional income through the sale of fruits, nuts, leaves, bark, resins, and other products.
3. Improving ecosystem services: MPTS contribute to carbon sequestration, soil conservation, water quality maintenance, and biodiversity enhancement.
4. Supporting livestock production: MPTS provide fodder, browse, and shade for livestock, improving their health and productivity.
5. Providing socio-cultural benefits: MPTS can serve as symbols of cultural identity, offer opportunities for recreation, and contribute to aesthetic values.

Selection of MPTS for Different Agroforestry Systems

When selecting MPTS for specific agroforestry systems, several factors should be considered:

1. Compatibility with associated crops and/or livestock: The chosen MPTS should not compete excessively with crops or livestock for resources such as light, water, and nutrients.
2. Site-specific suitability: The MPTS should be well-adapted to the local climate, soil, and topographical conditions.
3. Market potential: The products derived from the MPTS should have a viable market or potential for value addition.
4. Cultural preferences: The MPTS should align with local customs, traditions, and beliefs.
5. Environmental benefits: The MPTS should contribute to ecosystem services and sustainability.

MPTS of India

India is home to a rich diversity of MPTS, with several species being widely used in agroforestry systems. Some examples include:

1. Subabool (*Leucaena leucocephala*): A fast-growing, nitrogen-fixing tree, subabool is used for fodder, fuelwood, timber, and green manure.
2. Neem (*Azadirachta indica*): Neem is a versatile tree with numerous uses, including timber, fuelwood, fodder, medicine, and pest control.
3. Silver oak (*Grevillea robusta*): Primarily used for timber and fuelwood, silver oak also provides shade and improves soil fertility.
4. Indian mesquite (*Prosopis juliflora*): This tree is valued for its high-quality fuelwood, fodder, and nitrogen-fixing capabilities.

5. Gmelina (*Gmelina arborea*): Gmelina is a fast-growing, straight-grained timber tree used for construction, furniture, and paper production.
6. Sesbania (*Sesbania grandiflora*): Also known as hummingbird tree, sesbania is a leguminous tree with edible flowers, leaves, and pods, as well as green manure potential.

MPTS are integral to agroforestry systems, providing a wide range of benefits and services that enhance productivity, income, ecosystem health, and socio-cultural values. When selecting MPTS for specific agroforestry systems, it is essential to consider factors such as compatibility, site-specific suitability, market potential, cultural preferences, and environmental benefits. India's rich diversity of MPTS offers numerous opportunities for the development of sustainable and resilient agroforestry systems.

Different Agroforestry Systems, Subsystem, Practices, AFS Classification, AFS on Nature and Arrangement of Components

Different Agroforestry Systems

Agroforestry systems exhibit significant diversity and complexity across various regions of the world. Classifying these systems is crucial for evaluating their performance and developing effective improvement strategies. Several criteria can be used for classification, including system structure, function, socioeconomic factors, and ecological distribution. The terms "system," "sub-system," and "practices" are frequently used in Agroforestry and require clear definitions:

System

- ☆ A system refers to a collection of interconnected physical components that form a functional unit. For example, an ecosystem comprises living organisms and their non-living environment, which are inseparable and interrelated.
- ☆ In land use, a system represents a specific type of land use in a particular area, characterized by its biotechnical composition, arrangement, technical management level, and socio-economic features (*e.g.*, rice production system, plantation crop systems).

Sub-system

- ☆ A sub-system is a lower-level component within a system.
- ☆ It is a part of a system with a more restricted role, content, and complexity.
- ☆ A sub-system typically produces a defined "basic need" as its primary output, such as a food sub-system, an energy production sub-system, or a cash sub-system.

Practices

- ✰ Practices in Agroforestry refer to specific land management operations carried out on a farm or management unit.
- ✰ These practices are involved in establishing and maintaining an Agroforestry system (*e.g.*, alley cropping, boundary plantations of trees and shrubs, shelterbelts, and windbreaks).

Why Classify Agroforestry Systems?

- ✰ Classification provides a logical way to group the major factors that influence system production.
- ✰ It indicates how the system is managed.
- ✰ It offers flexibility for regrouping information.

Criteria/Basis for Agroforestry System Classification

Combe (1982) proposed 24 Agroforestry systems based on the association of trees with crops, pastures, or both; the trees' primary functions (production and protection); spatial arrangements (regular and irregular); and temporal associations (temporary and permanent).

The most practical criteria for classifying Agroforestry systems are the spatial and temporal arrangement of components, the relative importance and roles of those components, the production aims or outputs, and the socioeconomic features. These criteria align with the system's structure, function (output), socioeconomic nature, and ecological (environmental) spread, reflecting the main purposes of a classification scheme. Therefore, Agroforestry systems can be categorized according to these sets of criteria:

- ✰ Structural basis
- ✰ Functional basis
- ✰ Socioeconomic basis
- ✰ Ecological basis
- ✰ Vergera (1982) considered the relative allocation of land, trees, crops, pastures in various Agroforestry systems.

Tejwani (1987) suggested a classification which among the other things also took into account relative dominance of trees or crops/pastures

Classification

Classification of Agroforestry Systems (Nair, 1985)

Agroforestry systems can be classified based on their structural basis, which involves the composition of components, including the spatial arrangement of woody elements, vertical stratification of all components, and temporal

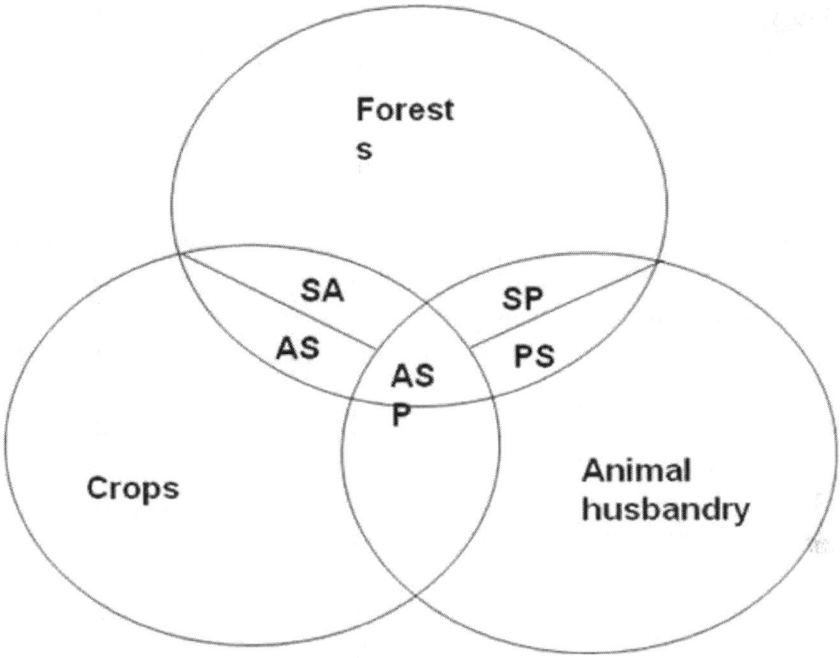

Figure 3.1: Agroforestry Systems on Basis of Relative Dominance of Components.

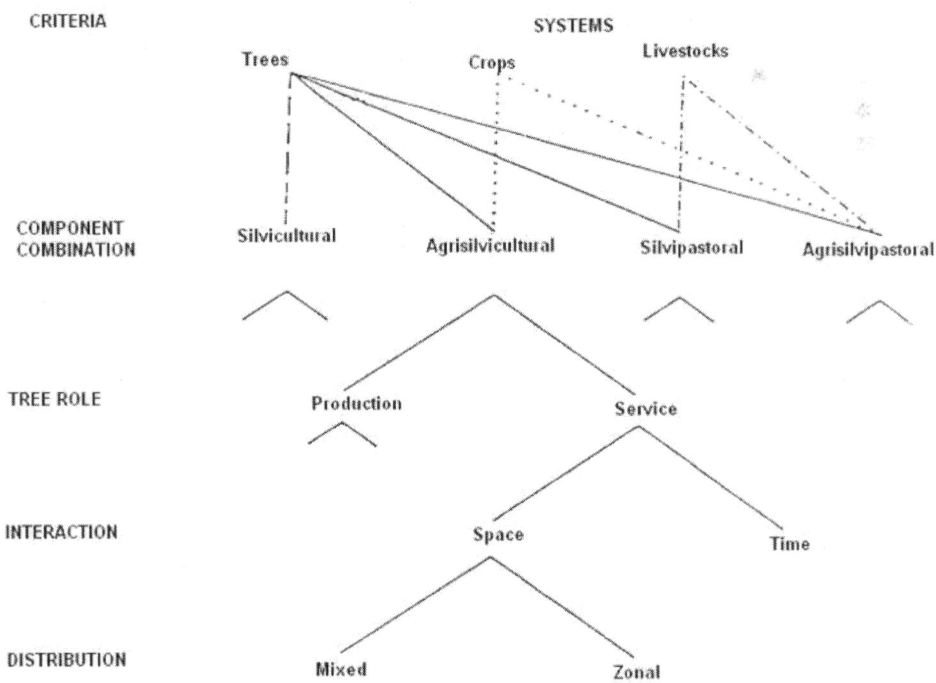

Figure 3.2: Agroforestry Systems Based on different Criteria.

sequencing of the different elements. As a result, Agroforestry systems can be categorized into two main groups based on structure.

A) Nature of Components

B) Arrangement of Components

A) Nature of Components

Agroforestry systems can be classified into the following categories based on the nature of their components:

- ☆ **Agri-silviculture systems (also known as silvi-agriculture or agro-silviculture):** These systems primarily integrate trees with agricultural crops.
- ☆ **Silvo pastoral systems (also known as silvi pastoral systems):** These systems combine trees with livestock or pasture.
- ☆ **Agro silvo pastoral systems (also known as agri silvi pastoral systems):** These systems incorporate trees, agricultural crops, and livestock together.
- ☆ **Other systems:** This category includes any additional combinations that do not fit neatly into the above groups.

Note: The nomenclature of an Agroforestry system depends on the primary importance of its components and the spatial prominence given to them. For example, in an agrisilviculture system, the agricultural crop is the dominant component, so it is prioritized in the name.

Agro silviculture systems have broad applicability and encompass the integration of various farming system components, such as vegetables, pulses, oilseed crops, and cereals, with tree crops. In contrast, agri silviculture systems are more limited, focusing only on the integration of cereals with tree crops.

Agrisilviculture/Silviagriculture/Agrosilviculture (Shifting Cultivation)

This system involves the conscious and deliberate use of land for the concurrent production of agricultural crops including tree, crops and forest crops. Based on the nature of the components this system can be grouped into various forms:

a) Improved fallow species in shifting cultivation
b) Taungya system
c) Multi species tree gardens
d) Alley cropping (Hedger winter-cropping)
e) Multi purpose trees and shrubs on farm lands
f) Crops combination with plantation crops

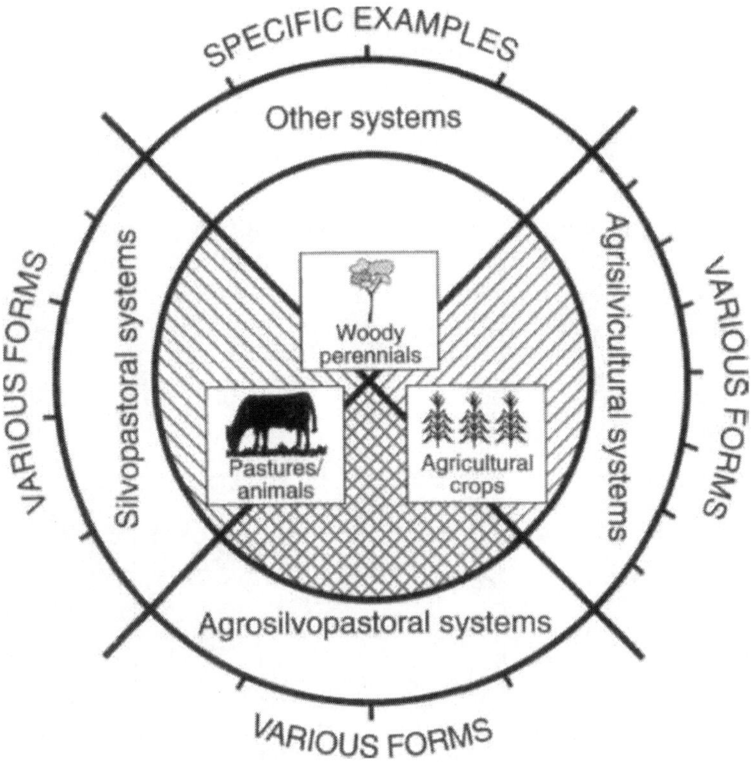

Figure 3.3: Classification of Agroforestry Systems on the Basis of Nature of Components.

g) Agroforestry for fuel wood production
h) Shelterbelts
i) Windbreaks
j) Soil conservation hedges *etc.*

a) Improved Fallow Species in Shifting Cultivation

Shifting Cultivation

- ☆ It is prevalent in many parts of Africa, Latin America, South-East Asia and Indian sub continent.
- ☆ In India, it is prevalent in the states of Assam, Meghalaya, Jharkhand, Manipur, Odisha, Nagaland, Chhattisgarh, Madhya Pradesh, Arunachal Pradesh, Andhra Pradesh, Mizoram, Tripura, Kerala, West Bengal, and Sikkim.
- ☆ It is known as 'jhuming' in North-east, 'khallu/kurwa' in Jharkhand and 'dahiya' or 'podo' in Orissa, Andhra Pradesh.

- In this system, forest patch is selected and cleared felled. The herbs, shrubs and twigs and branches (slashed vegetation) are burnt. Cultivation of crops is done for a few years until soil fertility declines. The site is than abandoned (fallow period) and new patch is selected for cultivation of crops. The site is again cultivated after giving rest for few years.
- Earlier the fallow cycle was of 20–30 year. However, due to increasing requirement for cultivation of land due to population pressure, fallow period has reduced from 25–30 years to 2–3 years which has broken down the resilience of ecosystem and the land is increasingly deteriorating. Thus now shifting cultivation has become source of ecological degradation, soil erosion and converting good forests into wastelands.

Figure 3.4: Shifting Cultivation.

Effect of Shifting Cultivation

- Deforestation and denudation of hill slopes-in secondary succession, area is occupied by weeds, useless shrubs *etc.*
- Soil erosion which leads to soil and nutrient losses, silting of reservoirs and streams, reduction in water-yield and landslips and landslides
- Shifting cultivation adversely affects cation exchange capacity and physical properties of soil. It leads to lowering of organic matter and lowering the total quantity of sesquioxides, iron, aluminum, calcium, potassium, phosphorus, *etc.*
- Increases soil pH and reducing microbial activity
- More weed growth and lower crop yield
- No opportunity for infrastructural development

Controlling Shifting Cultivation

- Motivate public for permanent agriculture by opening demonstration centers for improved agricultural practices, good quality seed, manuring, irrigation, weeding use of improved tools, terracing *etc.*
- Earning goodwill of local people: By engaging them in forest work and training them to undertake shifting cultivation on scientific lines.
- Arable land can be provided to the tribal for carrying out agriculture and also to settle in the area; a few schemes are being implemented under integrated tribal development programme
- **Legal measures:** on steep slopes, near to roadside *etc.*
- Using land according to its capability
- Provision of alternative management
- Development of animal husbandry and dairy farming
- Training of artisans and development of handicrafts
- Employment inforest works and other industries
- Providing communication facilities
- Providing economic assistance for houses and agriculture operations

Improved Fallow Species in Shifting Cultivation

- Fallows are cropland left without crops for periods ranging from one season to several years.
- The objective of improved fallow species in shifting cultivation is to recover depleted soil nutrients. Once the soil has recovered, crops are reintroduced for one or more season.
- The best species for the fallow system should induce good nitrogen fixation in the soil.
- The main aim of the fallow is to maintain or restore soil fertility and reduce erosion; some plants can be introduced primarily for their economic value.
- Plants included in improved fallows should be compatible with future crops, free of any negative physical or chemical effects on the soil and not in competition with the crops to be planted later on the same site.

b) Taungya System of Cultivation

The Taungya system is a form of Agroforestry that originated as a cost-effective method for establishing timber plantations but is now recognized as a legitimate Agroforestry practice.

Origin and History

- The term "taungya" (taung = hill, ya = cultivation) is Burmese, coined in Burma around 1850.
- Introduced to India by Brandis in 1890, with the first taungya plantations established in North Bengal in 1896.
- The system was also introduced to South Africa in 1887 and to Chittagong and Sylhet (now in Bangladesh) in 1870.
- In India, regular plantations started in North Bengal in 1911 for Sal and in 1912 for Teak. Uttar Pradesh adopted the system in 1923 for Sal plantations.

Current Status

- The Taungya system is still practiced in Kerala, West Bengal, Orissa, Karnataka, and the northeastern hill region of India.

Description

- Taungya is a modified form of shifting cultivation. Laborers are permitted to grow agricultural crops within an area designated for forest species planting.
- The process includes:
  - Land preparation
  - Tree planting
  - Cultivation of agricultural crops for 1-3 years (until shade becomes too dense)
  - Relocation to a new area to repeat the cycle.
- A variety of crops and trees are grown, depending on the local soil and climate.
- The primary purpose was to establish forest plantations, but the system has since been recognized as a valuable Agroforestry method.

Types of Taungya Systems

- **Departmental Taungya:** The forest department employs laborers on daily wages to cultivate agricultural crops alongside tree plantations, primarily to control weed growth.
- **Leased Taungya:** Plantation land is leased to individuals who offer the highest payment for cultivating crops for a set period while ensuring the care of the tree plantation.
- **Village Taungya:** Considered the most successful type. Villagers are settled within the forest to cultivate crops. Typically, each family manages 0.8 to 1.7 hectares of land for both trees and crops for 3 to 4 years.

Table 3.1: Trees and Crops Grown in Taungya

State	Tree Crop	Associated Agricultural Crops
U.P.	*Shorea robusta, Tectona grandis, Acacia catechu, Dalbergia sisso, Eucalyptus spp. Populus spp.*	Maize, paddy, sorghum, pigeon-pea, soyabean, wheat, barley, chick-pea, rape-seed and miscellaneous
Andhra Pradesh (AP)	*Anacardium occidentale, Tectona grandis, Bombax ceiba, Bamboo, Eucalyptus spp.*	Hillpaddy, groundnut, sweetpotato
Kerala	*Tectona grandis, Bombax ceiba Eucalyptus spp.*	Paddy, tapioca, ginger, turmeric, etc.
Assam	*Shore arobusta, Sassamica*	Paddy
TamilNadu	*Tectona grandis, Santalum album, Tamarindus indica, Acacia nilotica, Acacia mearnsii, Ceiba pentandra Cashew, Rubber, Bamboo*	Millet, pulses, groundnut, cotton
Andaman and Nicobar Islands	*Pterocarpus dalbergioides*	Sugar-cane, maize
Maharashtra	*Tectona grandis, Acacia nilotica*	Sunhemp, jute, mesta, sunflower, castor *etc.*
Tripura	*Shorea spp., Schima spp., Michelia spp.*	Paddy, maizeetc
West Bengal	*Tectona grandis, Shorea robusta, Schima wallichii, Cryptomeria japonica, Quercus spp., Michelia doltsopa*	Paddy, maize, millets, turmeric, ginger, lady's, finger, pineapple, sunhemp
Karnataka	*Tectona grandis, Santalum album, Cassia siamea*	Paddy, tapioca, *etc.*

Taungya System: Advantages and Disadvantages

The Taungya system is a form of Agroforestry where crops are grown alongside trees during the early stages of forest regeneration.

Advantages

- ☆ **Cost-Effective Reforestation:** Provides an inexpensive method for artificial forest regeneration.
- ☆ **Employment Opportunities:** Helps alleviate unemployment issues.
- ☆ **Maximized Land Use:** Facilitates optimal utilization of land resources.
- ☆ **Economical Plantation Establishment:** Offers a low-cost approach to establishing forest plantations.
- ☆ **Financial Benefits:** Can be highly profitable for forest departments.

- **Food Production:** Allows for the cultivation of food crops within forest areas.
- **Weed Control:** Helps eliminate unwanted weeds and climber growth.

Disadvantages

- **Soil Degradation:** Can lead to loss of soil fertility and increased soil exposure.
- **Disease Risk:** Potential for epidemics among crops or people.
- **Legal Complications:** May create legal challenges related to land use and rights.
- **Increased Erosion:** Increases the land's susceptibility to accelerated erosion.
- **Labor Exploitation:** Can be viewed as a form of exploiting human labor.

c) Multi-Species Tree Gardens and Alley Cropping

These are alternative Agroforestry systems.

Multi-Species Tree Gardens

- Involve the cultivation of diverse tree species in a mixed arrangement.
- Primarily aimed at producing food, fodder, and wood products for household consumption and commercial sale.

Alley Cropping (Hedgerow Intercropping)

- An Agroforestry technique where arable crops are grown in the alleys between rows of perennial trees or shrubs, often leguminous species.
- The trees are managed as hedgerows, with regular pruning. The pruned leaves and twigs are used as mulch in the crop alleys.
- **Benefits of Mulching:**
    - Reduces soil evaporation.
    - Suppresses weed growth.
    - Adds nutrients and organic matter to the topsoil.
- **Primary Goals:**
    - Maintain or increase crop yields by improving soil quality, microclimate, and weed control.
    - Provide additional tree products such as fuelwood, building poles, food, medicine, and fodder.

Layout of Alley Cropping Systems

- The arrangement and spacing of hedgerows and crops are determined by factors such as plant species, climate, slope, soil conditions, and the space needed for human movement.

- **Hedgerow Orientation:** Ideally, hedgerows should run east to west to maximize sunlight exposure for plants on both sides.
- **Spacing Recommendations:**
  - Row Spacing: Typically 4 to 8 meters between rows.
  - Tree Spacing: 25 cm to 2 meters between trees within rows.
  - Closer spacing is generally used in humid areas, while wider spacing is suitable for sub-humid or semi-arid regions.

Figure 3.5: Alley Cropping.

Characteristics of Species for Hedge Row Intercropping

Alley cropping typically involves the use of leguminous tree species to enhance soil fertility through biological nitrogen fixation. Therefore, an ideal tree or shrub species for alley cropping should possess the following characteristics:

Alley Cropping Considerations

Ideal tree and shrub species for alley cropping should possess the following characteristics:

- ✰ **Rapid regrowth:** Ability to quickly re-sprout after pruning, coppicing, pollarding, or lopping.
- ✰ **Deep root system:** A deep taproot system that efficiently extracts moisture and nutrients from lower soil layers, minimizing competition with shallow-rooted agricultural crops.
- ✰ **Manageable lateral roots:** Shallow lateral roots that can be easily pruned by ploughing along the hedgerow without causing significant harm to the plants.
- ✰ **Fast decomposition:** Rapid breakdown of leaf litter.
- ✰ **Nitrogen fixation:** Ideally, the species should be capable of nitrogen fixation.
- ✰ **Adaptability:** Trees and shrubs should be tolerant of a range of conditions.

Promising Alley Cropping Species

Examples of suitable species include: *Gliricidia sepium, Flemingia macrophylla, Leucaena, Calliandra calothyrsus, Erythrina subumbrans, Albizia saman, Pithecellobium dulce, Paraserianthes falcataria, Acacia spp., Paraserianthes falcataria,* and *Cajanus cajan.*

Advantages of Alley Cropping

- ✰ **Enhanced crop yields:** Improved crop performance due to nutrient enrichment and increased soil organic matter.
- ✰ **Reduced chemical fertilizer use:** Decreased reliance on synthetic fertilizers.
- ✰ **Improved soil health:** Enhancement of the soil's physical properties.
- ✰ **Erosion control:** Reduced soil loss from erosion.
- ✰ **Additional products:** Provision of supplemental resources like forage, firewood, or stakes (when using multipurpose leguminous trees).
- ✰ **Weed suppression:** Enhanced weed management.

e) Multipurpose Trees and Shrubs on Farmlands

This system involves scattering various multipurpose tree species randomly or systematically on field bunds. Key components include multipurpose trees, fruit trees, and common agricultural crops. The primary function is to produce diverse tree products, and also provides fencing and plot demarcation. Examples include: *Leucaena leucocephala, Acacia albida, Cassia siamea, Casuarina equisetifolia, Azadirachta indica, Acacia senegal, Cocos nucifera, etc.*

f) Crop Combinations with Plantation Crops (Intercropping)

Perennial trees and shrubs like coffee, tea, coconut, and cocoa are often integrated into intercropping systems in several ways:

- **Integrated multistory systems:** Combining multiple plantation crops in a layered arrangement.
- **Alternating crop arrangements:** Planting plantation crops in alternating patterns.
- **Shade trees:** Utilizing trees to provide shade for plantation crops.
- **Intercropping with annual crops:** Growing agricultural crops between rows of plantation crops.

Examples of Shade Tree Usage

- **Tea:** *Camilia sinensis* is grown under the shade of *Albizia chinensis, A. odoratissim, A. lebbek, A. procera, Acacia lenticularis, Derris robusta, Grevillea robusta, Acacia spp., Erythrina lithosperma,* and *Indigofera tesmanii.*
- **Coffee:** *Coffea arabica* benefits from temporary shade provided by *Erythrina lithosperma,* along with permanent shade trees such as *Ficus glomerata, F. nervosa, Albizia chinensis, A. lebbek, A moluccana, A. sumatrana, Dalbergia latifolia, Artocarpus integrifolius, Bischofia javanica,* and *Grevillea robusta.*
- **Cacao:** *Theobroma cacao* is cultivated under the shade of coconut, areca nut, and *Dipterocarpus macrocarpa* (in forested areas).
- **Black Pepper:** *Piper nigrum* is supported by *Erithrina indica, Garuga pinnata, Spondias, Mangifera, Gliricidia maculata,* and *Grevillea robusta.*
- **Cardamom:** Small cardamom (*Elettaria cardamomum*) and large cardamom (*Ammomum subulatum; A. aromaticum*) thrive under the temporary shade of *Mesopsis emini.* Large Cardamom is cultivated under the shade of natural forests and planted trees such as *Alnus nepalensis, Schima wallichii; Cinchona spp.; Lagerstroemia spp., Albizia lebbek; Castanopsis tribuloides; C. hystrix; C. indica; Terminalia myriocarpa; Bischofia javanica.*

g) Agroforestry for Fuelwood Production

This system involves interplanting various multipurpose fuelwood species on or around agricultural lands. The trees can serve as fencing, shelterbelts, and boundary markers. Common fuelwood species include: *Acacia nilotica, Albizia lebbek, Cassia siamea, Casuarina equisetifolia, Dalbergia sissoo, Prosopis juliflora, Eucalyptus tereticornis, etc.*

h) Shelterbelts

A shelterbelt is a wide belt of trees, shrubs, and grasses planted in rows perpendicular to prevailing winds. It deflects wind, reduces wind velocity, and protects cultivated areas from wind erosion and the drying effects of hot winds on the leeward side. A typical shelterbelt has a triangular cross-section, achieved by planting tall trees in the center, flanked by progressively shorter trees, tall shrubs, and then low spreading shrubs and grasses. Some penetrability is desirable to maximize the zone of influence and create a gradual reduction in wind speed. The width of the shelterbelt depends on local climate, wind velocity, and soil type. Shelterbelts should ideally be oriented at right angles to the prevailing wind. In areas with variable wind directions, they should be planted in quadrangles.

Height and Spacing

The height of a shelterbelt is crucial as it determines the distance of leeward protection. Taller trees provide a greater zone of influence. Spacing between shelterbelts is determined by the need to completely control wind erosion. If wind erosion has to be completely controlled, the second belt should be located a little before the place where the wind on the lee-ward side often first shelterbelt assumes damaging velocity.

Assuming a 20 per cent reduction in wind velocity as a measure of effectiveness, the protected zone extends up to 15-20 times the height of the belt. In Rajasthan, with a shelterbelt height of about 7.5 meters, a spacing of 75 meters (10 times the height) is recommended.

Length

The length of a shelterbelt is important to minimize eddies at the ends, which can increase wind velocity. Roads should not cross shelterbelts for this reason. In some countries, shelterbelts have been established across entire regions for maximum protection. For shorter shelterbelts to be effective, the minimum length should be 24 times its height.

Soil Preparation

Soil preparation should commence at least a year in advance to build up soil moisture reserves. This can be done mechanically or manually. Planting leguminous crops between rows of trees and shrubs in the initial years can improve soil fertility.

Shelterbelts: Species Selection, Characteristics, Establishment, and Benefits

Choice of Species

- ✰ The selection of tree species for shelterbelts is primarily determined by the local climate, soil conditions, and topography.

- Prioritizing local species is advantageous due to their inherent adaptability and ease of establishment.
- Exotic species can potentially be incorporated to enhance the overall effectiveness of the shelterbelt.

Characteristics of Suitable Tree Species

Ideal shelterbelt tree species should possess the following characteristics:

- **Non-exacting:** Able to thrive in a range of conditions.
- **Fast-growing:** Provides quick establishment and wind protection.
- **Wind-firm:** Resistant to wind damage.
- **Drought-resistant:** Tolerant of dry conditions.
- **Unpalatable to Animals:** Less susceptible to browsing damage.
- **Dense Crown and Low Branching:** Creates an effective wind barrier.
- **Persistent Foliage:** Retains leaves during periods when protection is most needed.
- **Multipurpose:** Provides economic benefits such as firewood, timber, and fodder.

Recommended Species

The following species are commonly used for shelterbelt creation:

- **Grasses:** *Cenchrus barbatus, Saccharum spontaneum, Saccharum munja, Panicum turgidum, Panicum antidotale*
- **Shrubs:** *Calotropis procera, Crotolaria burhia, Calligonum polygonoides, Clerodendron phlomoides, Cassia auriculata, Dodonaea viscosa, Jatropha curcas, Leptadenia spartivm, Agave spp., Sesbania aculeata.*
- **Small Trees:** *Acacia jacquemontii, Acacia leucophloea, Balanites aegyptiaca, Capparis aphylla, Salvadora oleoides.*
- **Trees:** *Acacia arabica, Acacia senegal, Acacia cyanophylla, Albizzia lebbek, Azadirachta indica, Dalbergia sissoo, Lannea coromendelica, Parkinsonia aculeata, Prosopis cineraria, Prosopis juliflora, Pongamia pinnata, Tecoma undulata, Tamarix articulata, Eucalyptus spp., Acacia tortilis.*

Plant Establishment

- **Nursery Propagation:** Starting seeds in polythene bags within a dedicated nursery is the preferred method to ensure healthy plant development and easier transplanting.
- **Watering:** Regular watering is essential for the first one to two years after planting.
- **Protection:** Adequate fencing is crucial to protect young plants from grazing animals.

Advantages of Shelterbelts

While further research is needed, existing studies (CAZRI, TNAU, and international research) demonstrate the following benefits:

- **Temperature Moderation:**
  - Shelterbelts buffer air and soil temperatures, reducing maximums and increasing minimums.
  - Daytime temperatures are typically cooler within the shelterbelt, and nighttime temperatures are warmer.
- **Increased Humidity:** Relative humidity can increase by 1-50 per cent in areas protected by shelterbelts.
- **Reduced Evapotranspiration:** Shelterbelts significantly lower evapotranspiration rates within their zone of influence.
- **Increased Soil Moisture:**
  - Soil moisture content is higher on the downwind (leeward) side of shelterbelts.
  - Shelterbelts promote water infiltration, contributing to groundwater recharge.
- **Wind Velocity Reduction and Erosion Control:**
  - Shelterbelts deflect wind upwards, significantly reducing wind velocity on the leeward side for a distance of 15-20 times the height of the trees.
  - Reduced wind velocity minimizes wind erosion.
- **Increased Crop Yields:**
  - Shelterbelts can increase agricultural and horticultural crop production.
  - Studies have shown significant yield increases in cotton, oats, and fruit orchards.
  - Fodder yield increases of 300-400 per cent have been reported.
- **Protection of Property:**
  - Shelterbelts act as barriers against shifting sands, protecting roads, railways, canals, and agricultural fields from sand deposition.

Other Agroforestry Techniques

- **Windbreaks:** Protective plantings, typically consisting of 2-3 rows of trees or shrubs spaced 0.5-2.5 meters apart, designed to shield gardens, farms, or fields from strong winds.
- **Soil Conservation Hedges:** Systems integrating multipurpose/fruit trees with agricultural species, primarily for soil conservation and tree product provision. Common species include: *Grevillea robusta,*

Acacia catechu, Pinus roxburghii, Acacia modesta, Prosopis juliflora, Alnus nepalensis, Leucaena leucocephala, etc.

- ☆ **Horti silviculture:** The integration of horticultural trees with timber trees on the same land unit to harvest fruits and timber concurrently. Timber trees are often planted on orchard bunds to act as windbreaks.
- ☆ **Horti silvopastoral:** A system that combines orchards with improved leguminous grasses to provide forage for livestock. Trees planted on the orchard bunds serve as windbreaks and provide multiple products.

QUESTION BANK

1. **What are Multi-Purpose Tree Species (MPTS)?**
   A. Trees used only for timber production
   B. Trees that provide multiple outputs and services like food, fodder, fuel, and timber
   C. Trees that grow only in forests
   D. Exotic species used for landscaping

Answer: B. Trees that provide multiple outputs and services like food, fodder, fuel, and timber

2. **Which of the following is NOT a typical product or service offered by MPTS?**
   A. Fuelwood
   B. Fiber
   C. Pesticides
   D. Fodder

Answer: C. Pesticides

3. **One of the most important criteria in selecting an MPTS for an agroforestry system is:**
   A. High water requirement
   B. Invasive growth habit
   C. Compatibility with crops and/or livestock
   D. Short lifespan

Answer: C. Compatibility with crops and/or livestock

4. **Which of the following is a well-known MPTS commonly used in Indian agroforestry systems?**
   A. Eucalyptus globulus
   B. Pinus wallichiana
   C. Azadirachta indica (Neem)
   D. Sequoia sempervirens

Answer: C. Azadirachta indica (Neem)

5. **What role do nitrogen-fixing MPTS play in agroforestry systems?**
   A. Increase soil acidity
   B. Reduce biodiversity
   C. Improve soil fertility through biological nitrogen fixation
   D. Reduce water infiltration

Answer: C. Improve soil fertility through biological nitrogen fixation

6. **Which MPTS is widely used in agroforestry for its ability to improve soil fertility and provide fodder, especially in semi-arid regions?**
   A. Acacia nilotica
   B. Populus deltoides
   C. Prosopis juliflora
   D. Leucaena leucocephala

Answer: D. Leucaena leucocephala

7. **MPTS are preferred in agroforestry because they:**
   A. Require no maintenance
   B. Are only ornamental
   C. Provide multiple benefits and enhance system resilience
   D. Compete aggressively with crops

Answer: C. Provide multiple benefits and enhance system resilience

8. **Which of the following characteristics should be avoided when selecting an MPTS for farmland?**
   A. Allelopathic effects on crops
   B. Deep root system
   C. Nitrogen-fixing ability
   D. Drought tolerance

Answer: A. Allelopathic effects on crops

9. **Gliricidia sepium is an example of MPTS known for:**
   A. Fiber production
   B. Ornamental value
   C. Fodder, green manure, fuelwood
   D. Essential oil extraction

Answer: C. Fodder, green manure, fuelwood

10. **Which of the following is not an essential consideration in selecting MPTS for a specific agro-climatic region?**
    A. Soil and moisture conditions
    B. Cultural compatibility
    C. Export potential
    D. Growth habit and canopy structure

Answer: C. Export potential

Chapter 4

Ecological Aspects of Agroforestry System, Tree-Crop Interaction: Competition, Nutrient Recycling

The Symbiotic Dance: Ecological Aspects of Agroforestry Systems

Agroforestry, the intentional integration of trees and shrubs with crops and/or livestock, isn't just a farming technique; it's a carefully orchestrated ecological system. By mimicking natural ecosystems, agroforestry harnesses the power of plant interactions to create resilient, productive, and environmentally beneficial landscapes. Understanding the ecological aspects of agroforestry, particularly tree-crop interactions like competition and nutrient recycling, is crucial for successful implementation and maximizing its potential.

A Tapestry of Interactions: Understanding Tree-Crop Dynamics

The heart of agroforestry lies in the dynamic interplay between trees and crops. These interactions can range from beneficial symbiosis to detrimental competition, depending on the species chosen, their spatial arrangement, and management practices. Recognizing and managing these interactions is paramount.

Competition: A Necessary Evil, Tamed by Design

Competition for resources like light, water, and nutrients is an inevitable facet of any multi-species system. In agroforestry, understanding and mitigating this competition is key to optimizing productivity.

- **Light Competition:** Trees, by virtue of their height, can shade crops, reducing photosynthetic efficiency. Strategically pruning tree canopies, selecting low-growing tree varieties, and orienting rows to maximize sunlight penetration are effective strategies to minimize light competition and ensure adequate light reaches the crops.
- **Water Competition:** In water-scarce regions, competition for water can be a significant constraint. Deep-rooted trees can access water resources unavailable to shallow-rooted crops, but they can also deplete soil moisture. Choosing drought-tolerant species, employing water-harvesting techniques, and strategically placing trees to capture runoff can help mitigate water competition.
- **Nutrient Competition:** Similar to water, trees and crops both require essential nutrients for growth. Excessive competition for nutrients can lead to nutrient deficiencies and reduced yields. Soil testing and targeted fertilization, coupled with the strategic placement of trees to optimize nutrient uptake, can alleviate nutrient competition.

However, it's important to remember that competition can also be beneficial. Strategic competition can suppress weeds naturally, reducing the reliance on herbicides. Choosing trees with allelopathic properties, meaning they release chemicals that inhibit the growth of other plants, can further enhance weed suppression.

Nutrient Recycling: Nature's Fertilizer Factory

One of the most significant ecological benefits of agroforestry is its capacity for efficient nutrient cycling. Trees act as nutrient pumps, drawing nutrients from deeper soil layers that are inaccessible to crops.

- **Litterfall and Decomposition:** As trees shed their leaves, branches, and fruits, this organic matter, known as litterfall, decomposes and releases nutrients back into the soil. This natural fertilization process reduces the need for synthetic fertilizers, improving soil health and minimizing environmental pollution.
- **Nitrogen Fixation:** Leguminous trees, in symbiosis with nitrogen-fixing bacteria in their roots, can convert atmospheric nitrogen into a form usable by plants. This biological nitrogen fixation enriches the soil with a vital plant nutrient, reducing the demand for nitrogen fertilizers.
- **Mycorrhizal Associations:** Many trees form symbiotic relationships with mycorrhizal fungi, which enhance nutrient uptake, particularly phosphorus. These fungi extend the root system's reach, facilitating access to nutrients and improving overall plant health.
- **Reduced Soil Erosion:** Tree roots bind the soil together, reducing erosion and nutrient loss from runoff. This is particularly important in sloping landscapes, where erosion can be a major problem.

Beyond Competition and Recycling: Other Ecological Benefits

The ecological advantages of agroforestry extend beyond competition management and nutrient cycling. Agroforestry systems also:

- **Enhance Biodiversity:** Agroforestry provides diverse habitats for wildlife, supporting a wider range of species compared to monoculture farming systems.
- **Improve Soil Health:** Increased organic matter content from tree litter improves soil structure, water infiltration, and water-holding capacity.
- **Sequester Carbon:** Trees absorb carbon dioxide from the atmosphere, mitigating climate change.
- **Provide Shade and Shelter:** Trees offer shade and shelter for crops and livestock, buffering them from extreme temperatures and wind.

Conclusion: Orchestrating an Ecological Symphony

Agroforestry is more than just planting trees alongside crops. It's about understanding the intricate web of interactions between different species and harnessing these interactions to create a sustainable and productive farming system. Carefully selecting compatible species, managing competition effectively, and leveraging the natural nutrient cycling processes are crucial for realizing the full ecological and economic potential of agroforestry. By embracing the principles of ecological design, we can transform agricultural landscapes into resilient ecosystems that benefit both people and the planet.

Different Agroforestry Systems, Subsystem, Practices, AFS Classification, AFS on Nature and Arrangement of Components

Silvipastoral Systems (Trees + Pasture and/or Animals)

A silvopastoral system integrates woody plants (trees and shrubs) with pastureland and/or livestock. These systems are valuable, especially in drier regions, for providing wood and fodder throughout the year. The trees and shrubs can serve multiple purposes:

- **Fodder Production:** Providing feed for livestock.
- **Timber, Fuelwood, and Fruit:** Producing marketable products.
- **Soil Improvement:** Enhancing soil health and fertility.

Three main categories of silvopastoral systems exist:

A. Protein Banks

- **Description:** Multipurpose trees, rich in protein, are planted on or around farmlands and rangelands.

- **Purpose:** To provide a "cut and carry" fodder source, supplementing livestock diets during periods of fodder scarcity (*e.g.*, winter).
- **Examples:** *Grewia optiva, Bauhinia variegata, Morus alba, Artocarpus spp., Anogeissus latifolia, Cordia dichotoma, Dalbergia sissoo, Eutralobium saman, Zizyphus jujube, etc.*

B. Live Fences of Fodder Trees and Hedges

- **Description:** Fodder trees and shrubs are planted as living fences around properties.
- **Purpose:** Protecting property from stray animals and other biotic influences.
- **Examples:** *Sesbania grandiflora, Gliricidia sepium, Erythrina abyssinica, Euphorbia spp., Acacia spp., etc.*

C. Trees and Shrubs on Pastureland

- **Description:** Trees and shrubs are scattered irregularly or arranged in a systematic pattern across pastureland.

Agri-silvopastoral/Agro-silvopastoral Systems (Crops + Trees + Grasses/Animals)

These systems combine crops, trees, and grasses/animals. Two main subgroups are recognized:

A. Home Gardens

- **Description:** A deliberate integration of trees, crops, and animals on a single land unit, often found within the compounds of individual houses, using spatial and temporal arrangements.
- **Characteristics:**
  - Highly productive and sustainable.
  - Typical of high rainfall areas in South and Southeast Asia.
  - Common in Southern India (Kerala, Tamil Nadu) and Northeastern states (Tripura, Assam, West Bengal, Andaman and Nicobar Islands).
  - Also known as multilayered Agroforestry System (AFS).
  - Homestead area typically ranges from 0.2-0.5 hectares.
  - **Structure:** Tall timber trees form the upper layer, followed by fruit trees, shrubs, and shade-loving vegetables at ground level.
  - Trees provide timber, fruits, and support for climbers like pepper, cucurbits, clove, yam, and sweet potato.
  - Pineapple is a common fruit crop.

- **Examples (Hills):** *Grewia optiva, Ficus glomerata, Juglans regia,* and *Punica granatum.*
- **Examples (Rural Areas):** Fruit trees and commercial species like *Acacia* and *Neem.*
- **Livestock:** Cattle and poultry are common components.
- **Forage:** Species like *Stylo, Guinea grass, Guatemala, Napier,* and *Setaria cephalis* variety Kazungula may be included.

B. Woody Hedges for Browsing, Green Manuring, Mulching, and Soil Conservation

- **Description:** Fast-growing, coppicing fodder shrubs and trees are planted as hedges.
- **Purpose:** Providing browse for animals, green manure, mulch, and soil conservation benefits. The primary goals are food/fodder/fuelwood production and soil conservation.
- **Examples:** *Erythrina spp., Leucaena leucocephala,* and *Sesbania spp., etc.*

Other Specified Systems

i) Apiculture with Trees

- **Description:** Nectar and pollen-rich trees/shrubs are planted on farm bunds, sometimes in combination with agriculture or oilseed crops.
- **Examples:** *Mangifera indica, Vitex negundo, Melia azedarach, Azadirachta indica, Prunus salicina, Prunus armeniaca, Rubus ellipticus, Eucalyptus spp., Callistemon lanceolatus, Berberis lycium, Toona ciliata, etc.*
- **Purpose:** Honey production.

ii) Aqua-forestry

- **Description:** Common in coastal regions (especially along the Andhra coast), integrating fish/prawn cultivation in saline water with tree cultivation on pond bunds.
- **Benefits:** Trees provide litter-feed for aquaculture and generates income for farmers. Fish culture in mangroves provides nutrition and breeding grounds for aquatic life.
- Well-balanced systems may include goatry, poultry, duck-farming, turtles, and fishes in small ponds, maximizing resource use.
- Leguminous tree leaves (*e.g., Gliricidia sepium, Leucaena, Moringa oleifera, Acacia nilotica*) can serve as fish feed.
- The area is enclosed by earth embankments and a system of ridges and canals inside helps collect rainwater and grow tree species.

iii) Multipurpose Woodlots

- **Description:** Location-specific multipurpose tree species (MPTs) are grown in mixed or separate plantings.
- **Purpose:** Production of wood, fodder, soil protection, and soil reclamation.

Agroforestry Systems: Arrangement of Components

Arrangement of components prioritizes plants. Even with animals, management follows a plan (*e.g.*, rotational grazing) that favors plant needs. Plant arrangements consider space and time.

- **Spatial Arrangement:**
  - **Dense Mixed Stands:** E.g., home gardens.
  - **Sparse Mixed Stands:** E.g., trees in pastures.
  - **Zonal/Strip Arrangement:** Species are laid out in zones or strips. Includes microzonal and macrozonal arrangements.
  - **Hedge Intercropping:** A common example of zonal patterns.
  - **Boundary Planting:** Trees on plot edges for fruits, fodder, fuelwood, fencing, soil protection, and windbreaks.
  - **Macrozonal arrangement:** Can be laid to sole cropping system but the interactions association of different components can be used criteria to decide like limits between zonal AF and sole crop (component) plots

Temporal Arrangement in Agroforestry Systems

The temporal arrangement of plants within Agroforestry systems describes how different species occupy the land over time. Here are several common patterns:

- **Coincident:** Woody and non-woody components occupy the same space simultaneously (*e.g.*, coffee plants growing under shade trees, or pasture grasses beneath shade trees).
- **Concomitant:** Woody and non-woody components co-exist for a portion of their lifecycles (*e.g.*, the Taungya system, where crops are grown alongside young trees).
- **Intermittent (Space-Dominated):** Annual crops are grown alongside perennial crops (*e.g.*, paddy rice cultivated with coconut trees). Space is the primary organizing factor.
- **Interpolated (Space and Time-Dominated):** Different components occupy the same space at different times (*e.g.*, a home garden, where various species are planted and harvested sequentially). Both space and time are key considerations.

- **Overlapping:** A system where the lifecycles of component crops overlap in time (*e.g.*, Black pepper vines growing on Rubber trees).
- **Separate (Time-Dominated):** Components occupy the land at distinct times (*e.g.*, using improved fallow species in shifting cultivation, where different species occupy a space in distinct temporal phases).

Functional Basis of Agroforestry Systems

The functional basis of an Agroforestry system refers to the primary roles or purposes it serves. These can be broadly categorized as:

- **Productive Functions:**
  - Food
  - Fodder
  - Fuelwood
  - Other wood products
  - Other marketable products
- **Protective Functions:**
  - Windbreak
  - Shelterbelt
  - Soil Conservation
  - Moisture Conservation
  - Soil Improvement
  - Shade (for crops, animals, and people)

Socioeconomic Basis of Agroforestry Systems

The socioeconomic basis considers the level of management input, the intensity or scale of management, and the commercial goals associated with the system. It can be broadly classified into:

- **Commercial Agroforestry Systems:** Primarily focused on producing a profitable output (*e.g.*, commercial tree plantations with intercropped food crops).
- **Intermediate Agroforestry Systems:** A balance between commercial and subsistence goals.
- **Subsistence Agroforestry Systems:** Primarily designed to meet the basic needs of the owner/occupant and their family. Selling surplus produce is secondary.

Ecological Basis of Agroforestry Systems

The ecological basis considers the environmental conditions and ecological suitability of the systems. The idea is that certain types of Agroforestry systems are better suited for specific ecological conditions. For example, different sets of systems might be developed for arid and semi-arid regions, tropical highlands, or humid lowland tropics.

QUESTION BANK

1. **Which of the following is an ecological benefit of agroforestry systems?**
   A. Increased use of chemical fertilizers
   B. Enhanced biodiversity and soil conservation
   C. Reduced organic matter in soil
   D. Elimination of native species

Answer: B. Enhanced biodiversity and soil conservation

2. **Tree-crop competition in agroforestry primarily occurs for:**
   A. Temperature and wind
   B. Light, water, and nutrients
   C. Pesticides and insecticides
   D. Market and labor

Answer: B. Light, water, and nutrients

3. **Which type of root system in trees is preferred to minimize competition with crops?**
   A. Surface spreading roots
   B. Shallow and fibrous roots
   C. Deep tap root system
   D. Aerial roots

Answer: C. Deep tap root system

4. **How can competition between trees and crops in agroforestry systems be reduced?**
   A. By planting only fast-growing trees
   B. By monocropping only
   C. Through proper spatial arrangement and species selection
   D. By removing leaf litter regularly

Answer: C. Through proper spatial arrangement and species selection

5. **Nutrient cycling in agroforestry is enhanced primarily through:**
   A. Leaching
   B. Leaf litter decomposition and root turnover
   C. Herbicide application
   D. Irrigation

Answer: B. Leaf litter decomposition and root turnover

6. **Trees in agroforestry systems can act as a "nutrient pump" by:**
   A. Reducing microbial activity
   B. Absorbing surface water
   C. Bringing nutrients from deeper soil layers to the surface
   D. Suppressing nitrogen fixation

Answer: C. Bringing nutrients from deeper soil layers to the surface

7. **One negative ecological impact of poor agroforestry design is:**
   A. Increased pollinator activity
   B. Better carbon sequestration
   C. Excessive competition leading to reduced crop yield
   D. Improved microclimate

Answer: C. Excessive competition leading to reduced crop yield

8. **Allelopathy in tree-crop interactions refers to:**
   A. Mutual shading between species
   B. Symbiotic nitrogen fixation
   C. Release of chemicals by trees that inhibit crop growth
   D. Nutrient recycling through root nodules

Answer: C. Release of chemicals by trees that inhibit crop growth

9. **Which of the following trees is known for minimal competition and good nutrient recycling in agroforestry systems?**
   A. Eucalyptus spp.
   B. Leucaena leucocephala
   C. Prosopis juliflora
   D. Pinus roxburghii

Answer: B. Leucaena leucocephala

10. **A key ecological principle in designing agroforestry systems is to:**
    A. Maximize tree density to suppress weeds
    B. Use species with similar resource requirements
    C. Promote complementarity between species to reduce competition
    D. Focus only on fast-growing exotics

Answer: C. Promote complementarity between species to reduce competition

Chapter 5

Traditional Agroforestry as a Viable Choice to Conserve Agro-biodiversity of India

Traditional Agroforestry: A Time-Tested Solution for Conserving India's Agro-biodiversity

India, a land of diverse climates and rich agricultural heritage, boasts a remarkable wealth of agro-biodiversity. This biodiversity, encompassing the variety of crops, livestock, and associated flora and fauna, is crucial for food security, livelihood sustenance, and resilience to climate change. However, modern agricultural practices often prioritize monoculture and high-yielding varieties, threatening this invaluable resource. In this context, traditional agroforestry systems emerge as a viable and time-tested approach to conserve and enhance India's precious agro-biodiversity.

What is Traditional Agroforestry?

Traditional agroforestry, deeply rooted in indigenous knowledge and practices, involves the deliberate integration of trees and shrubs with crops and/or livestock. These systems are not mere haphazard combinations; they are carefully designed and managed to optimize resource use, minimize risk, and enhance overall productivity while respecting local ecological conditions. Across India, various forms of traditional agroforestry exist, each adapted to specific agro-climatic zones and cultural contexts. Examples include:

- **Home gardens:** Multi-layered systems surrounding homesteads, integrating fruit trees, vegetables, spices, and livestock. These gardens serve as living gene banks, preserving a wide range of local varieties.

- **Taungya Systems:** Combining tree crops with agricultural crops during the early stages of tree establishment, benefiting both forestry and agriculture.
- **Shifting Cultivation (Jhum):** A cyclical system involving clearing and burning forest patches for cultivation, followed by a fallow period for regeneration. While often demonized, traditional Jhum practices, when managed sustainably, can contribute to biodiversity conservation.

Why Agroforestry for Agro-biodiversity Conservation?

The benefits of traditional agroforestry for agro-biodiversity conservation are manifold:

- **Habitat Provision:** Trees and shrubs provide diverse habitats for beneficial insects, pollinators, birds, and other wildlife, promoting ecological balance and pest control.
- **Soil Health Improvement:** Agroforestry practices enhance soil fertility through nitrogen fixation by leguminous trees, organic matter accumulation, and improved water infiltration. Healthy soils are essential for supporting diverse plant life.
- **Microclimate Regulation:** Trees provide shade, reducing evapotranspiration and creating a more favorable microclimate for crops, particularly in arid and semi-arid regions.
- **Crop Diversification:** Agroforestry encourages the cultivation of a variety of crops, reducing dependence on single varieties and enhancing resilience to pests, diseases, and climate variability.
- **Preservation of Traditional Knowledge:** These systems are often managed based on intricate indigenous knowledge regarding species selection, planting techniques, and resource management, safeguarding this valuable intangible heritage.
- **Genetic Resource Conservation:** Agroforestry systems often act as in-situ conservation sites for neglected and underutilized species (NUS), contributing to the preservation of genetic resources.

Challenges and Opportunities

Despite the numerous benefits, traditional agroforestry faces challenges:

- **Land Tenure Insecurity:** Lack of clear land ownership can discourage long-term investment in agroforestry systems.
- **Lack of Awareness:** Many farmers are unaware of the potential benefits of agroforestry or lack the knowledge and skills to manage these systems effectively.
- **Market Access:** Limited access to markets for agroforestry products can hinder the economic viability of these systems.

- **Policy Support:** Inadequate policy support and lack of integration of agroforestry into agricultural development programs can limit its adoption.

However, opportunities exist to overcome these challenges and promote the widespread adoption of traditional agroforestry:

- **Strengthening Land Tenure:** Clear and secure land tenure rights are essential to incenti*viz*.e investment in agroforestry.
- **Capacity Building:** Training programs for farmers on agroforestry techniques, species selection, and market development can enhance their skills and knowledge.
- **Market Development:** Promoting the value of agroforestry products through certification schemes and direct marketing initiatives can improve market access.
- **Policy Integration:** Integrating agroforestry into agricultural development policies, providing incentives, and promoting research on agroforestry systems can accelerate its adoption.

Traditional agroforestry is not just a farming practice; it is a conservation strategy that safeguards India's invaluable agro-biodiversity. By integrating trees and shrubs with crops and livestock, these systems create resilient and sustainable landscapes that support both livelihoods and the environment. Recognizing the potential of traditional agroforestry and addressing the challenges it faces is crucial for securing India's agricultural future and preserving its rich agro-biodiversity for generations to come. Investing in research, promoting awareness, and providing policy support for traditional agroforestry will be key to unlocking its full potential as a viable and sustainable choice for the future of Indian agriculture.

Introduction

- India is one of the world's 2nd largest populated country.
- India has huge human population of 125 crore.
- Most of the population (75 per cent) residing in rural area which totally depends upon forest to meet out their energy requirement.
- The demand for fuelwood in India is increasing day by day.
- India's current firewood consumption is more than 133 million tonnes; most of it is being used in cooking. To cook 1 kg of food, 1.2 kg of firewood is required.
- It clearly indicates that India should produce more wood than food if it is to be cooked before it is consumed. The electricity can also be generated by dried wood.

- According to estimate, 400 million tonnes of cattle dung equivalent to about 60 million tonnes of fuelwood are burnt annually in our country.
- If this much quantity of cattle dung is incorporated into the soil, then it could increase soil productivity.
- Similarly, fuelwood is the most significant reason for tree cutting.
- To save forests from degradation, fuel wood tree growing should become part of agriculture through agroforestry in blocks in order to meet out their demands of fuel wood and improve the microclimate by means of saving trees in natural forests.
- An energy plantation is one that is grown purely for plant material for their fuel than for fibre content.

Criteria of Tree Species Planted for Energy Plantation

- Tree species should be fast growing with high photosynthetic efficiency which results into high yields.
- Tree species should have high coppicing and pollarding capacity.
- Tree species selected to energy plantation should be conical or cylindrical in shape.
- Tree species should have wood of high calorific value, high wood density, dry weight and burns without sparks or toxic smoke.
- Tree species should be able to tolerate incidence of insects, pests and diseases.
- Tree species should have ability in them to reduce transpiration loss in arid areas.
- Tree species should have ability to fix nitrogen, if possible, that can improve soil fertility without having much competition with main crop for soil moisture, sunlight, *etc.*
- Tree species should be multiple in nature.

Suitable Species for Firewood/Fuelwood/Energy Plantation for Different Regions

Tropical Dry Region

Acacia catechu, Acacia modesta, Acacia nilotica, Acacia senegal, Acacia tortilis, Anogeis suspendula, Albizia lebbek, Azadirachta indica, Cassia siamea, Cordia rothii, Dalbergia sissoo, Emblica officinalis, Eucalyptus camaldulensis, Erythrina superb, Gmelina arborea, Parkinsonia aculeate, Peltophorum ferrugineum, Pongamia pinnata, Prosopis cineraria, Prosopis juliflora, Tamarindus indica, Tamarix troupii, Tecomella undulata, Zizyphus maurtiana etc.

Tropical Humid Region

Adina cordifolia, Acacia auriculiformis, Acacia catechu, Acacia nilotica, Albizia procera, Azadirachta indica, Cassia siamea, Casuarina equisetifolia, Dalbergia sissoo, Dendrocalamus strictus, Ficus spp., Eucalyptus spp., Kydia calycina, Leucaena leucocephala, Madhuca indica, Melia azedarach, Morus alba, Salix tetrasperma, Syzygium cuminii, Tamarindus indica, Trema nudiflora, Gliricidia sepium, Gmelina arborea.

Sub-Tropical Region

Acacia catechu, Acacia melanoxylon, Acacia nilotica, Aesculus indica, Ailanthus excelsa, Celtis australis, Grevillea robusta, Michelia champaca, Populus deltoides, Populus nigra, Robinia pseudoacacia, Salix alba, Toona ciliata.

Temperate Climate

Acer spp., Aesculus indica, Alnus nepalensis, Alnus nitida, Celtis australis, Populus ciliata, Quercus semecarpifolia, Salix alba, Toona serrata.

- ☆ The direct use of firewood in densely populated area should be avoided as it causes environmental pollution.
- ☆ Some firewoods on burning give toxic and irritating smoke, and foul odour.
- ☆ The firewood may be converted into charcoal which is more efficient.

Charcoal

- ☆ Charcoal is an ideal smokeless fuel for cooking. 1 kg of charcoal has a replacement value of 2.38 kg of firewood or more.
- ☆ The combustion efficiency of charcoal is about 28 percent.
- ☆ Thus, conversion of firewood into charcoal for use as a fuel will be better than firewood as such.
- ☆ Charcoal is also useful as a reductant in electrometallurgical industries.
- ☆ Manufacture of calcium carbide, carbon-disulphide and active carbon.
- ☆ It does not contain sulphur.

Important Tree Species for Charcoal Making

Acacia nilotica, Adina cordifolia, Anogeissus latifolia, Casuarina equisetifolia, Pinus roxburghii, Quercus leucotrichophora, Quercus semecarpifolia, Tamarindus indica, Terminalia arjuna, Terminalia bellerica, Terminalia chebula, Terminalia catappa.

Shrubs for Energy Plantation

Atlantia monophylla, Crewia latifolia, Clerodendron inerme, Dodonaea viscosa, Jatropha glandulifera, Jatropha curcas, Tecoma gracilis, Ipomoea camea etc.

- Besides firewood and charcoal, plants also provide exudates and extractives.
- Such plant species are energy rich and may be exploited as renewable sources of energy.
- These species are known as 'petro-crops', since they can serve as substitutes for or supplements to petrochemicals.

Extractive Plants

Based on exudates and extractives, plants are classified as those bearing:

1. Latex
2. Vegetable oil and waxes
3. Resins
4. Essential oils
5. Tannins and phenolic compounds bearing plants

Latex Yielding Plant Species

- Plant species yielding latex belong to Family *Apocynaceae, Asclepiadaceae, Euphorbiaceae, Moraceae,* and *Sapotaceae.*
- Potential petro-crops are: *Euphorbia antisyphilitica, E. tirucalli, E. lathyris, Pedilanthes tithymaloides, Calotropis procera, Asclepias curassavica, Parthenium argentatum.*

Vegetable Oils

- Vegetable oils have great potential to be used as liquid fuel or as a source of hydrocarbons.
- Some of them can be mixed in diesel.
- The non-edible seed bearing oil tree species can be cultivated on poor, marginal and wastelands.

Important species are:

Antinodaphe hookeri, Aleurites triloba, Anacardium occidentale, Aphanamixis polystachya, Azadirachta indica, Calophyllum inophyllum, Cocos nucifera, Croton tiglium, Garcinia indica, Hydnocarpus wightiana, Jatropha curcas, Madhuca indica, Madhuca longifolia, Melia azedarach, Mesua ferrea, Mimusops elengi, Pongamia pinnata, Pittosporum resiniferum, Ricinus communis, Salvadora oleoides, Sapium sebiferum, Schleichera oleosa, Semecarpus anacardium, Shorea robusta, Simmondsia chinensis, Strychnos nux-vomica, Vateria indica.

Resins

- Resins are collected mainly from members of family *Pinaceae.*

- These are volatile oils (turpentine) and non-volatile resins (rosin).
- The resins are main source for synthetic rubber and other polymers.
- Turpines are highly combustible and they can be used in various formulations of fuel for automobiles.

Calorific Value

- The amount of heat produced when 1 g of fuel is completely burnt in excess of air or oxygen.
- If one gram of carbon is burned completely, it produces about 30,000 J or 30 KJ/g of heat.
- Therefore, the calorific value of carbon is 30 KJ/g and fuel having high calorific value is regarded as good fuel.
- CV of hydrogen is 150 KJ/g. However, it is not commonly used fuel because of highly combustible nature and difficulty in its handling.

Table 5.1: A Few Species Used in Energy Plantations with their Respective Calorific Value and Specific Gravity

Sl.No.	Species	Sp. Gravity	Calorific Value Kcal/kg
1.	*Acacia auriculiformis*	0.60-0.78	4800-4900
2.	*Acacia catechu*	1.00	5142-5244
3.	*Acacia dealbata*	0.70-0.85	3500-4000
4.	*Acacia leucophloea*	0.78	4899-4886
5.	*Acacia mearnsii*	0.70-0.85	3500-4000
6.	*Acacia nilotica*	0.67-0.68	4800-4950
7.	*Acacia senegal*	-	3200
8.	*Acacia tortilis*	-	4400
9.	*Adina cordifolia*	-	3855
10.	*Aegle marmelos*	0.91	4495
11.	*Albizia lebbek*	0.55-0.64	5163-5166
12.	*Albizia odoratissima*	0.73	5131-5266
13.	*Albizia procera*	0.68	4870-4865
14.	*Alnus nepalensis*	0.32-0.37	4600
15.	*Anogeissus latifolia*	0.94	4948
16.	*Anogeissus pendula*	0.94	4900
17.	*Anthocephalus cadamba*	0.94-0.53	4800
18.	*Artocarpus heterophyllus*	0.51	5318
19.	*Azadirachta indica*	0.75	-
20.	*Baringtonia acutangula*	0.58	5078

Sl.No.	Species	Sp. Gravity	Calorific Value Kcal/kg
21.	*Bauhinia retusa*	0.72	5027
22.	*Bauhini avariegata*	-	4800
23.	*Butea monosperma*	0.54	4909
24.	*Bischofia javanica*	0.74	5162
25.	*Cajanus cajan*	-	4594
26.	*Cassia siamea*	0.60-0.80	-
27.	*Casuarina equisetifolia*	0.80-1.2	4950
28.	*Cedrela toona*	0.57	5113-5168
29.	*Chloroxylons wietenia*	-	4759
30.	*Dalbergia sissoo*	0.75-0.80	4908-5181
31.	*Diospyros melanoxylon*	0.79-0.87	4957-5030
32.	*Diospyros montana*	0.70-0.80	5125
33.	*Dodonaea viscosa*	1.20-1.28	5035-4939
34.	*Emblica offcinalis*	0.70-0.80	5200
35.	*Eucalyptus camaldulensis*	0.6	4800
36.	*Eucalyptus globulus*	0.80-1.00	4800
37.	*Eucalyptus grandis*	0.40-0.70	4900
38.	*Eucalyptus tereticornis*	0.70	4800
39.	*Gmelina arborea*	0.42-0.64	4763-4800
40.	*Greville arobusta*	0.57	4904-4914
41.	*Grewia spp.*	0.67	5292
42.	*Hardwickia binata*	1.08	4891-4952
43.	*Holoptelia integrifolia*	0.63	5228
44.	*Lannea coromandelica*	0.55	4933
45.	*Leucaena leucocephala*	0.55-0.70	4200-4600
46.	*Madhuca longifolia*	0.56	5043-5156
47.	*Mangifera indica*	0.58	4610
48.	*Melia azedarach*	0.56	5043-5176
49.	*Morus alba*	0.63	4371-4773
50.	*Michelia champaca*	0.45	5068
51.	*Ougeinia oojeinensis*	0.85	5178
52.	*Pithecellobiumdulce*	0.64	5177-5600
53.	*Pongamia pinnata*	0.75	4600
54.	*Populus euphratica*	0.48	5008-5019
55.	*Prosopis chilensis*	0.80-0.92	5000-5500
56.	*Prosopis cineraria*	0.77-0.94	5000
57.	*Prosopis juliflora*	0.70	4800

Sl.No.	Species	Sp. Gravity	Calorific Value Kcal/kg
58.	*Pterocarpus marsupium*	0.79	4904-5141
59.	*Pterygota alata*	0.25-0.62	5160
60.	*Quercus leucotrichophora*	0.74	4633
61.	*Schleichera oleosa*	0.91-1.08	4928-4950
62.	*Sesbania grandiflora*	0.55	4407
63.	*Shorea robusta*	0.68-0.82	5095-5433
64.	*Syzygium cuminii*	0.67-0.78	4834
65.	*Tamarindus indica*	0.91-1.28	4909-4969
66.	*Tamarixa phylla*	0.60-0.75	4835
67.	*Tectona grandis*	0.55-0.70	4989-5535
68.	*Terminalia alata*	0.71-0.94	5047-5373
69.	*Terminalia arjuna*	0.74-0.82	5030-5128
70.	*Terminalia chebula*	0.77	3967
71.	*Trema orientalis*	0.48	3095
72.	*Xylia xylocarpa*	0.92	4975-5044
73.	*Zizyphus mauritiana*	0.93	4900

Advantages of Energy Plantations

- ☆ Emit little or no sulphur and less nitrogen dioxide than fossil fuel
- ☆ Helps in rehabilitation of degraded lands
- ☆ Provide rural employment
- ☆ Alive and active growing forest and other plant biomass absorb the greenhouse gas in quantities broadly equivalent to amount emitted when plant material decay or burned. They are thus called as "Carbon neutral" fuel sources
- ☆ Growing energy crops creates a "carbon sink" which includes storing carbon underground through the tree root system
- ☆ Lower energy cost per unit area as lower inputs are required as compared

QUESTION BANK

1. **Traditional agroforestry systems in India are best described as:**
   A. Recently introduced farming methods
   B. Monoculture systems focused on timber production
   C. Indigenous land-use practices integrating trees, crops, and livestock
   D. Systems based only on exotic tree plantations

Answer: C. Indigenous land-use practices integrating trees, crops, and livestock

2. **Which of the following is a major ecological benefit of traditional agroforestry systems?**
   A. Increased pesticide usage
   B. Soil degradation
   C. Conservation of agrobiodiversity and genetic resources
   D. Reduction in organic matter content

Answer: C. Conservation of agrobiodiversity and genetic resources

3. **The homegarden agroforestry system, commonly found in Kerala and northeastern India, is an example of:**
   A. Commercial forestry
   B. Mixed-species traditional agroforestry
   C. Plantation agriculture
   D. Industrial agroforestry

Answer: B. Mixed-species traditional agroforestry

4. **Which Indian community is known for its traditional Zabo system that combines agriculture, forestry, and animal husbandry?**
   A. Bhils B. Nagas
   C. Gonds D. Meenas

Answer: B. Nagas

5. **Traditional agroforestry systems help in the conservation of agrobiodiversity by:**
   A. Eliminating native species
   B. Encouraging monocultures
   C. Preserving traditional crop varieties and wild relatives
   D. Promoting chemical inputs

Answer: C. Preserving traditional crop varieties and wild relatives

6. **Which of the following traditional systems is practiced in arid Rajasthan and integrates *Prosopis cineraria* with crops like pearl millet and legumes?**

   A. Taungya
   B. Khadin
   C. Agri-silvipastoral system
   D. Parkland agroforestry

Answer: D. Parkland agroforestry

7. **Why are traditional agroforestry systems considered resilient to climate change?**

   A. Use of genetically modified species
   B. Monoculture-based planting
   C. High diversity of species and ecological functions
   D. Dependence on high external inputs

Answer: C. High diversity of species and ecological functions

8. **Which traditional agroforestry system of the Western Ghats involves tree-based spice farming (*e.g.*, black pepper, cardamom)?**

   A. Taungya system
   B. Shifting cultivation
   C. Multistoried homegardens
   D. Silvopasture

Answer: C. Multistoried homegardens

9. **Traditional agroforestry systems in tribal areas often serve which additional purpose beyond food and fiber?**

   A. Industrial processing
   B. Cultural and religious significance
   C. Urban development
   D. Export-oriented production

Answer: B. Cultural and religious significance

10. **A key strategy to conserve India's agrobiodiversity through agroforestry is to:**

   A. Promote large-scale deforestation
   B. Replace all native species with fast-growing exotics
   C. Revitalize and document traditional agroforestry knowledge systems
   D. Focus solely on timber extraction

Answer: C. Revitalize and document traditional agroforestry knowledge systems

Chapter 6

Management of Agroforestry System

Managing for Success: Navigating the Complexities of Agroforestry Systems

Agroforestry, the intentional integration of trees and shrubs into agricultural lands, holds immense promise for sustainable land management. However, harnessing its full potential requires careful planning and diligent management. Unlike monoculture farming, agroforestry presents a dynamic, multi-layered system demanding a holistic approach to ensure both ecological and economic success.

This article delves into the intricacies of managing agroforestry systems, highlighting key considerations and strategies for optimizing their performance.

1. Planning and Design: The Foundation for Effective Management

Effective management begins long before planting. Thorough planning and design are crucial for establishing a system that meets specific goals. Key considerations include:

- ☆ **Objectives:** Clearly define the goals of the agroforestry system. Are you aiming for timber production, soil conservation, enhanced biodiversity, fruit production, or a combination of benefits?
- ☆ **Species Selection:** Carefully choose tree and crop species that are compatible, complement each other, and thrive in the local environment. Consider factors like growth rates, light requirements, nutrient needs, and potential allelopathic interactions.

- **Spatial Arrangement:** Determine the optimal arrangement of trees and crops. This could involve alley cropping, windbreaks, hedgerows, or silvopasture systems. Factors like sunlight penetration, root competition, and accessibility for management operations should be taken into account.
- **Market Analysis:** Understand the market demand for both tree and crop products to ensure profitability and sustainability.

2. Establishment and Early Care: Setting the Stage for Long-Term Success

The initial establishment phase is critical for the long-term success of the agroforestry system. Key practices include:

- **Site Preparation:** Prepare the site adequately by clearing weeds, improving soil drainage, and ensuring proper soil fertility.
- **Planting:** Plant trees and crops at the appropriate spacing and time of year. Use high-quality seedlings and seeds.
- **Weed Control:** Implement effective weed control measures to prevent competition with young trees and crops.
- **Protection:** Protect young trees from browsing animals and pests using fencing, tree guards, or other appropriate methods.
- **Irrigation and Fertilization:** Provide adequate irrigation and fertilization, especially during the initial establishment phase, to promote healthy growth.

3. On-going Management Practices: Nurturing a Thriving Ecosystem

Once the agroforestry system is established, ongoing management is essential for maintaining productivity and achieving desired outcomes. This includes:

- **Pruning and Thinning:** Prune trees regularly to improve light penetration, enhance fruit production, and promote desirable tree shapes. Thinning may also be necessary to reduce competition and improve the growth of remaining trees.
- **Crop Management:** Manage crops using appropriate agronomic practices, including fertilization, irrigation, pest and disease control, and harvesting.
- **Soil Management:** Implement soil conservation practices to prevent erosion and maintain soil fertility. This could include cover cropping, mulching, and contour farming.

- **Pest and Disease Management:** Monitor the system for pests and diseases and implement appropriate control measures, prioritizing integrated pest management (IPM) strategies to minimize environmental impact.
- **Nutrient Management:** Monitor soil nutrient levels and apply fertilizers as needed to ensure adequate nutrition for both trees and crops. Consider using organic fertilizers and soil amendments to improve soil health.
- **Water Management:** Optimize water use through efficient irrigation techniques and water harvesting strategies, particularly in arid and semi-arid regions.

4. Harvesting and Marketing: Realizing the Economic Benefits

Proper harvesting and marketing strategies are essential for maximizing the economic benefits of agroforestry. This includes:

- **Harvesting:** Harvest crops and tree products at the appropriate time to ensure optimal quality and market value.
- **Processing and Storage:** Process and store products properly to prevent spoilage and maintain their quality.
- **Marketing:** Develop effective marketing strategies to reach potential buyers and secure favorable prices. Consider direct marketing, farmers' markets, and collaborations with local businesses.

5. Monitoring and Evaluation: Adapting for Continuous Improvement

Regular monitoring and evaluation are essential for tracking the performance of the agroforestry system and identifying areas for improvement. This includes:

- **Data Collection:** Collect data on tree growth, crop yields, soil health, biodiversity, and economic performance.
- **Analysis:** Analyze the data to assess the effectiveness of management practices and identify any problems or challenges.
- **Adaptation:** Adapt management practices as needed to address identified problems and improve the overall performance of the system.

Challenges in Agroforestry Management

While agroforestry offers numerous benefits, it also presents unique management challenges:

- **Complexity:** Managing a multi-species system is inherently more complex than managing monoculture farming.

- **Competition:** Competition between trees and crops for resources like sunlight, water, and nutrients can be a significant challenge.
- **Labor Requirements:** Agroforestry systems often require more labor than conventional farming systems.
- **Knowledge Gap:** Many farmers lack the knowledge and skills needed to effectively manage agroforestry systems.
- **Policy and Institutional Support:** Supportive policies and institutional frameworks are often lacking, hindering the adoption and scaling up of agroforestry practices.

Managing agroforestry systems effectively requires a holistic and adaptive approach. By carefully planning and designing the system, implementing appropriate management practices, and continuously monitoring and evaluating its performance, farmers can harness the full potential of agroforestry to achieve both ecological and economic benefits. Overcoming the challenges through knowledge sharing, supportive policies, and innovative management techniques will be crucial for promoting the widespread adoption of agroforestry as a sustainable land management practice. The future of sustainable agriculture may well depend on our ability to effectively manage these complex and rewarding systems.

QUESTION BANK

1. **Which of the following is the primary objective of managing an agroforestry system?**

   A. Maximizing tree height

   B. Maximizing crop monoculture

   C. Sustainable production of food, fodder, fuel, and timber

   D. Promoting deforestation

Answer: C. Sustainable production of food, fodder, fuel, and timber

2. **The most important factor to consider while designing an agroforestry system is:**

   A. Tree species only   B. Soil fertility only

   C. Climatic condition & land capability   D. Crop yield only

Answer: C. Climatic condition and land capability

3. **Which of the following is a silvicultural practice used in agroforestry management?**

   A. Pruning   B. Broadcasting

   C. Ploughing   D. Spraying herbicides

Answer: A. Pruning

4. **Thinning in agroforestry systems is done to:**

   A. Increase pest infestation   B. Encourage weed growth

   C. Reduce competition among trees   D. Remove healthy trees

Answer: C. Reduce competition among trees

5. **Lopping of trees in agroforestry systems is mainly done to:**

   A. Promote water logging   B. Reduce soil fertility

   C. Increase light availability for crops   D. Increase tree height only

Answer: C. Increase light availability for crops

6. **What is the key management practice to reduce competition between trees and crops in agroforestry?**

   A. Increasing irrigation   B. Application of herbicides

   C. Proper spacing and canopy management   D. Monoculture planting

Answer: C. Proper spacing and canopy management

7. Which tree management technique is used to harvest leaves, branches, or fruits without cutting down the entire tree?

A. Felling
B. Girdling
C. Coppicing
D. Lopping

Answer: D. Lopping

8. In agroforestry, rotational management refers to:

A. Planting only one crop throughout the year
B. Shifting land use from trees to crops and back periodically
C. Cutting all trees at once
D. Use of chemical fertilizers

Answer: B. Shifting land use from trees to crops and back periodically

9. The best time for planting trees in an agroforestry system in India is:

A. Winter season
B. Post-monsoon season
C. Pre-monsoon or early monsoon season
D. Summer season

Answer: C. Pre-monsoon or early monsoon season

10. Which of the following practices is not typically a part of agroforestry system management?

A. Pollarding
B. Mulching
C. Clear felling
D. Coppicing

Answer: C. Clear felling

Chapter 7

Role of Agroforestry in Soil and Water Conservation

The Root of Resilience: How Agroforestry Conserves Soil and Water

In a world grappling with climate change and increasing pressure on natural resources, sustainable land management practices are more crucial than ever. Among these, agroforestry stands out as a powerful tool for conserving two of our most vital resources: soil and water. By integrating trees and shrubs into agricultural landscapes, agroforestry systems offer a multifaceted approach to environmental stewardship, fostering both productivity and ecological well-being.

Understanding Agroforestry: A Symbiotic Relationship

Agroforestry is not a new concept; it's an age-old practice rooted in indigenous knowledge systems. It involves the intentional integration of trees and shrubs with crops and/or livestock on the same land management unit. This deliberate combination creates a symbiotic relationship, where each component benefits the other, leading to enhanced ecological and economic outcomes.

The Soil's Best Friend: Agroforestry for Soil Conservation

Soil erosion is a major threat to agricultural productivity, leading to land degradation and reduced yields. Agroforestry combats this issue through several mechanisms:

- ☆ **Root Systems: The Anchor:** Tree roots act as a natural anchor, binding the soil particles together and preventing them from being washed away by rain or blown away by wind. This is particularly important on sloping lands where erosion is more pronounced.

- ☆ **Reduced Runoff:** The canopy cover provided by trees intercepts rainfall, reducing the impact of raindrops on the soil surface. This slows down runoff, allowing more water to infiltrate the soil and reducing the risk of erosion.
- ☆ **Improved Soil Structure:** Tree roots create macro-pores in the soil, improving aeration and drainage. The decomposition of leaf litter and other organic matter from trees also enhances soil structure, making it more resistant to erosion.
- ☆ **Nutrient Cycling:** Trees can access nutrients from deeper soil layers, which are then deposited on the soil surface through leaf litter. This improves soil fertility and reduces the need for synthetic fertilizers, ultimately contributing to soil health.

Water Warriors: Agroforestry for Water Conservation

Water scarcity is an increasing concern in many regions. Agroforestry plays a crucial role in conserving water resources by:

- ☆ **Increased Infiltration:** As mentioned earlier, agroforestry improves soil structure, leading to increased water infiltration. This replenishes groundwater reservoirs and reduces surface runoff.
- ☆ **Reduced Evapotranspiration:** While trees do transpire, the overall evapotranspiration rate in agroforestry systems can be lower compared to monoculture agriculture. This is because the canopy provides shade, reducing soil temperature and evaporation.
- ☆ **Water Quality Improvement:** Trees act as natural filters, trapping sediments and pollutants from agricultural runoff. This improves water quality in streams and rivers, benefiting both human and ecological health.
- ☆ **Rainwater Harvesting:** Agroforestry can be strategically designed to enhance rainwater harvesting, collecting and storing rainwater for later use in irrigation or other purposes.

Beyond Conservation: Multiple Benefits of Agroforestry

The advantages of agroforestry extend far beyond soil and water conservation. It also provides:

- ☆ **Increased Biodiversity:** Agroforestry systems create diverse habitats for wildlife, promoting biodiversity and ecosystem services.
- ☆ **Carbon Sequestration:** Trees absorb atmospheric carbon dioxide, mitigating climate change.

- ☆ **Economic Diversification:** Agroforestry provides farmers with multiple income streams from crops, livestock, and tree products.
- ☆ **Improved Livelihoods:** By enhancing soil fertility, water availability, and economic opportunities, agroforestry contributes to improved livelihoods for rural communities.

The Path Forward: Scaling Up Agroforestry for a Sustainable Future

Agroforestry offers a powerful pathway towards sustainable land management. To realize its full potential, it is essential to:

- ☆ **Promote awareness and education:** Educate farmers and policymakers about the benefits of agroforestry.
- ☆ **Provide technical support:** Offer training and technical assistance to farmers on how to design and manage agroforestry systems.
- ☆ **Develop supportive policies:** Implement policies that incenti*viz*.e agroforestry practices, such as subsidies and tax breaks.
- ☆ **Invest in research and development:** Conduct research to improve agroforestry techniques and adapt them to different agro-ecological zones.

By embracing agroforestry, we can build more resilient and sustainable agricultural systems that conserve our precious soil and water resources, ensuring a healthier planet for future generations. The root of resilience lies in the symbiotic relationship between trees, crops, and people, a relationship that agroforestry embodies and promotes.

QUESTION BANK

1. Which of the following is a key function of agroforestry in soil conservation?

A. Increase soil erosion

B. Increase runoff

C. Reduce wind and water erosion

D. Compete with crops for nutrients

Answer: C. Reduce wind and water erosion

2. Tree roots in agroforestry systems help in:

A. Competing for nutrients

B. Binding soil particles and reducing erosion

C. Increasing soil salinity

D. Promoting soil cracking

Answer: B. Binding soil particles and reducing erosion

3. The most effective agroforestry system for controlling wind erosion in arid areas is:

A. Agrisilviculture

B. Hortisilviculture

C. Shelterbelts and windbreaks

D. Home gardens

Answer: C. Shelterbelts and windbreaks

4. Trees improve soil organic matter through:

A. Pesticide application

B. Leaf litter and root biomass

C. Acid rain absorption

D. Fire management

Answer: B. Leaf litter and root biomass

5. Which agroforestry practice is best for stabilizing terraces and slopes?

A. Home garden

B. Live hedgerows

C. Aquaforestry

D. Monocropping

Answer: B. Live hedgerows

6. How does agroforestry contribute to water conservation?

A. By reducing infiltration

B. By increasing runoff

C. By improving infiltration and reducing evaporation losses

D. By preventing rainfall

Answer: C. By improving infiltration and reducing evaporation losses

7. **Trees in agroforestry systems reduce the impact of raindrops on soil surface, which is called:**

A. Evaporation control

B. Splash erosion control

C. Infiltration reduction

D. Hydraulic conductivity

Answer: B. Splash erosion control

8. **Deep-rooted trees improve water availability by:**

A. Blocking water movement

B. Bringing nutrients and moisture from deeper soil layers

C. Increasing soil compaction

D. Enhancing surface runoff

Answer: B. Bringing nutrients and moisture from deeper soil layers

9. **Which agroforestry intervention enhances soil moisture retention and groundwater recharge?**

A. Strip cropping

B. Contour planting of trees

C. Mono-cropping

D. Deep tillage

Answer: B. Contour planting of trees

10. **One of the major benefits of agroforestry for degraded lands is:**

A. Complete replacement of crops

B. Restoration of soil health and productivity

C. Reduced biodiversity

D. Elimination of native vegetation

Answer: B. Restoration of soil health and productivity

Chapter 8

Windbreak and Shelterbelt–Definition, Objectives

Windbreaks and Shelterbelts: Nature's Barriers for a Better Environment

Wind, while a valuable resource for energy and dispersal, can also be a destructive force. It can erode soil, damage crops, stress livestock, and increase energy consumption in buildings. Fortunately, we can harness the power of nature to mitigate these negative impacts through the strategic planting of **windbreaks** and **shelterbelts**. These linear plantings of trees and shrubs offer a multitude of benefits, contributing to a healthier environment and a more sustainable future.

So, what exactly are windbreaks and shelterbelts? And what are the objectives behind their implementation? Let's delve into the details:

Definition

While often used interchangeably, there's a subtle distinction between the two:

- **Windbreak:** Generally refers to a single row or a few rows of trees or shrubs designed to reduce wind speed and its impact in a localized area. They are often used around homes, gardens, or small agricultural plots.
- **Shelterbelt:** Refers to a more extensive, multi-row planting of trees and shrubs, often covering a larger area and providing broader protection,

such as for a whole field or farmstead. Shelterbelts tend to have a more complex structure, incorporating a variety of species for enhanced effectiveness and ecological benefits.

In essence, both serve the same primary function: to obstruct and deflect wind, creating a zone of reduced wind speed on their leeward (downwind) side.

Objectives

The objectives behind implementing windbreaks and shelterbelts are numerous and varied, depending on the specific location and need. However, some key objectives are common across most applications:

- **Soil Erosion Control:** One of the most crucial benefits is the reduction of wind erosion. By slowing down wind speeds, windbreaks prevent valuable topsoil from being carried away, particularly in agricultural areas and exposed landscapes. This helps maintain soil fertility and prevents dust storms.
- **Crop Protection:** Windbreaks can significantly improve crop yields by providing a more stable and favorable microclimate. They reduce wind damage to plants, prevent lodging (falling over), and improve pollination. They can also help trap snow, providing valuable moisture for crops in arid regions.
- **Livestock Protection:** In agricultural settings, shelterbelts provide valuable shelter for livestock from harsh weather conditions, such as strong winds, extreme temperatures, and heavy snowfall. This reduces stress on animals, improves their health and productivity, and lowers mortality rates.
- **Wildlife Habitat:** Windbreaks and shelterbelts create valuable habitat for a variety of wildlife species. They provide food, shelter, nesting sites, and travel corridors, enhancing biodiversity in agricultural landscapes.
- **Energy Conservation:** Strategically placed windbreaks around buildings can significantly reduce heating and cooling costs by reducing heat loss in winter and providing shade in summer.
- **Snowdrift Control:** In areas prone to heavy snowfall, windbreaks can be used to trap snow, preventing it from accumulating on roads, driveways, and buildings. This reduces the need for snow removal and improves safety.
- **Aesthetic Enhancement:** Windbreaks and shelterbelts can enhance the beauty of the landscape, providing visual interest and creating a more pleasing environment. They can also serve as screens, providing privacy and noise reduction.

Windbreaks and shelterbelts are powerful, nature-based solutions with a wide range of benefits for agriculture, the environment, and human well-being. By understanding their definition and objectives, we can effectively design and implement these valuable systems to create more sustainable and resilient landscapes. From protecting our precious soil to providing shelter for livestock and conserving energy, windbreaks and shelterbelts are a crucial tool in our efforts to combat the negative effects of wind and create a healthier future for all.

QUESTION BANK

1. **What is the primary purpose of windbreaks and shelterbelts in agroforestry?**
   A. Increase wind velocity
   B. Enhance desertification
   C. Reduce wind speed and protect crops
   D. Attract insects to fields

Answer: C. Reduce wind speed and protect crops

2. **Shelterbelts are typically planted:**
   A. Randomly across the field
   B. Parallel to slope direction
   C. Perpendicular to the prevailing wind direction
   D. Around water bodies only

Answer: C. Perpendicular to the prevailing wind direction

3. **Which of the following is not an effect of shelterbelts?**
   A. Reduction in wind erosion
   B. Decrease in crop yields
   C. Moderation of microclimate
   D. Protection of soil moisture

Answer: B. Decrease in crop yields

4. **An ideal shelterbelt should be composed of:**
   A. Only tall trees
   B. A mixture of trees, shrubs, and grasses
   C. Only grasses
   D. Only thorny bushes

Answer: B. A mixture of trees, shrubs, and grasses

5. **The effective protection zone on the leeward side of a windbreak is approximately:**
   A. 2–5 times the tree height
   B. 5–10 times the tree height
   C. 20–30 times the tree height
   D. 50 times the tree height

Answer: C. 20–30 times the tree height

6. **Which of the following tree species is commonly used in windbreaks in arid and semi-arid regions of India?**
   A. Mangifera indica
   B. Acacia nilotica
   C. Eucalyptus tereticornis
   D. Casuarina equisetifolia

Answer: D. Casuarina equisetifolia

7. **Windbreaks can help in reducing crop lodging by:**
   A. Increasing rainfall
   B. Blocking sunlight
   C. Slowing down high wind speed
   D. Increasing wind velocity

Answer: C. Slowing down high wind speed

8. **One key environmental benefit of shelterbelts is:**
   A. Enhanced carbon sequestration
   B. Increased soil erosion
   C. Depletion of groundwater
   D. Reduction in biodiversity

Answer: A. Enhanced carbon sequestration

9. **Which arrangement gives the best wind protection in a shelterbelt?**
   A. Trees planted in a single row
   B. Trees planted randomly
   C. Multi-row planting with different species and heights
   D. Only grass vegetation

Answer: C. Multi-row planting with different species and heights

10. **Windbreaks are also effective in:**
    A. Increasing pest incidence
    B. Reducing evaporation from soil and water surfaces
    C. Reducing soil fertility
    D. Preventing tree growth

Answer: B. Reducing evaporation from soil and water surfaces

Chapter 9

Socio-economic Aspects of Agroforestry System

The socio-economic aspects of agroforestry systems play a vital role in determining their adoption, sustainability, and overall impact on rural livelihoods. Agroforestry provides multiple benefits to farmers, including diversified income sources from timber, fuelwood, fruits, fodder, and non-timber forest products, thereby reducing dependence on a single crop and enhancing economic resilience. It also helps in ensuring food security by integrating trees with crops and livestock, which leads to more stable and productive systems. From a social perspective, agroforestry can promote community participation, equitable resource sharing, and gender inclusion, particularly when women are involved in managing home gardens and processing agroforestry products. Additionally, agroforestry contributes to employment generation, especially in rural and tribal areas, through nursery raising, planting, harvesting, and value addition of tree-based products. Land use efficiency increases under agroforestry, making better use of marginal and degraded lands, thus providing a livelihood option for landless or smallholder farmers. Moreover, agroforestry practices often align with traditional knowledge and cultural values of local communities, facilitating easier adoption and integration into existing farming systems. However, the socio-economic success of agroforestry also depends on supportive policies, access to markets, availability of extension services, and credit facilities. In summary, agroforestry not only enhances ecological sustainability but also strengthens socio-economic development at the household and community levels.

The Socio-Economic Aspects of Agroforestry Systems: A Sustainable Path to Prosperity

Agroforestry, a land use management system that combines trees, crops, and/or animals on the same piece of land, has gained significant attention in recent years due to its potential to address socio-economic and environmental challenges. By integrating various components, agroforestry systems offer a unique opportunity to improve the livelihoods of smallholder farmers, promote biodiversity conservation, and contribute to climate change mitigation and adaptation. This article explores the socio-economic aspects of agroforestry systems and their role in fostering sustainable development.

1. Improved Livelihoods and Food Security

Agroforestry systems can provide smallholder farmers with a diverse range of products, including food, fodder, fuelwood, timber, and non-timber forest products (NTFPs), thereby enhancing their income and food security. The integration of trees into agricultural landscapes not only increases productivity but also reduces the vulnerability of farmers to external shocks, such as crop failures, market fluctuations, and climate change. Moreover, agroforestry systems can serve as a safety net for resource-poor farmers, enabling them to meet their basic needs and invest in other income-generating activities.

2. Empowerment of Women and Marginalized Groups

Women and marginalized groups often bear the brunt of poverty, food insecurity, and limited access to resources. Agroforestry systems can help address these challenges by providing opportunities for income generation, skill development, and capacity building. Trees and their products, such as fruits, nuts, and medicinal plants, can be managed and marketed by women, thereby enhancing their economic empowerment and decision-making power within households and communities. Furthermore, agroforestry practices, such as boundary planting and home gardens, can be promoted as income-generating activities for women and other marginalized groups, contributing to their social and economic well-being.

3. Biodiversity Conservation and Ecosystem Services

Agroforestry systems can contribute to biodiversity conservation and the provision of ecosystem services, such as soil fertility enhancement, water regulation, and pest control. The integration of trees into agricultural landscapes can promote habitat heterogeneity, which in turn supports a diverse range of species, including pollinators, beneficial insects, and wildlife. Moreover, agroforestry practices, such as alley cropping and silvopasture, can help maintain soil health, reduce erosion, and enhance carbon sequestration, thereby mitigating climate change and promoting adaptation.

4. Market Opportunities and Value Chains

Agroforestry systems can provide farmers with access to new markets and value chains, thereby enhancing their income and livelihoods. Trees and their products, such as fruits, nuts, and timber, can be marketed through various channels, including local markets, processing industries, and export markets. Moreover, the integration of trees into agricultural landscapes can create opportunities for the development of new value chains, such as ecotourism, bioprospecting, and carbon sequestration.

5. Policy and Institutional Support

The success of agroforestry systems in improving socio-economic outcomes depends on supportive policies and institutions. Governments, development agencies, and civil society organizations can play a critical role in promoting agroforestry practices, providing technical assistance, and facilitating access to markets and finance. Moreover, the recognition of customary tenure rights and the promotion of community-based natural resource management can help ensure the sustainability of agroforestry systems and their contributions to socio-economic development.

Agroforestry systems offer a promising path to sustainable development, providing opportunities for improved livelihoods, biodiversity conservation, and climate change mitigation and adaptation. By integrating trees into agricultural landscapes, smallholder farmers can diversify their sources of income, enhance their food security, and contribute to the conservation of natural resources. Moreover, agroforestry systems can empower women and marginalized groups, create new market opportunities, and foster sustainable value chains. To realize these benefits, it is essential to promote supportive policies and institutions, provide technical assistance, and facilitate access to markets and finance. By doing so, agroforestry systems can contribute to a greener, more inclusive, and resilient future for all.

QUESTION BANK

1. **Which of the following is a primary socio-economic benefit of agroforestry systems?**
   A. Increased soil erosion
   B. Enhanced rural livelihoods and income diversification
   C. Reduced biodiversity
   D. Increased dependency on chemical fertilizers

Answer: B. Enhanced rural livelihoods and income diversification

2. **Agroforestry contributes to poverty alleviation primarily by:**
   A. Increasing monoculture farming
   B. Providing multiple sources of income throughout the year
   C. Reducing crop production
   D. Increasing the cost of inputs

Answer: B. Providing multiple sources of income throughout the year

3. **Which group benefits the most from agroforestry practices in rural areas?**
   A. Urban industrial workers
   B. Small and marginal farmers
   C. Large-scale plantation owners only
   D. Importers of agricultural products

Answer: B. Small and marginal farmers

4. **Agroforestry improves socio-economic status by:**
   A. Decreasing employment opportunities
   B. Increasing availability of fuelwood, fodder, and timber locally
   C. Increasing land degradation
   D. Restricting crop diversity

Answer: B. Increasing availability of fuelwood, fodder, and timber locally

5. **A major social benefit of agroforestry systems is:**
   A. Reduction of social conflicts over resources
   B. Displacement of local communities

C. Increase in migration from rural to urban areas

D. Exclusivity of resources for elite farmers

Answer: A. Reduction of social conflicts over resources

6. **Which factor limits the adoption of agroforestry in some rural areas?**

A. Abundance of agricultural land

B. Lack of awareness and technical knowledge

C. Government incentives and subsidies

D. High market demand for agroforestry products

Answer: B. Lack of awareness and technical knowledge

7. **Agroforestry systems provide employment opportunities primarily in:**

A. Urban construction industries

B. Rural areas through nursery management, planting, and harvesting

C. IT sector jobs

D. Export-import businesses

Answer: B. Rural areas through nursery management, planting, and harvesting

8. **How does agroforestry impact women in rural communities?**

A. Limits their access to resources

B. Enhances their involvement and income generation opportunities

C. Excludes them from decision-making processes

D. Increases their workload without benefits

Answer: B. Enhances their involvement and income generation opportunities

9. **Which of the following is an economic challenge faced by agroforestry farmers?**

A. High initial investment cost and delayed returns

B. Immediate high profits

C. Zero risk of crop failure

D. Constant government subsidies

Answer: A. High initial investment cost and delayed returns

10. Agroforestry contributes to sustainable development by:

A. Promoting monoculture farming

B. Balancing ecological health and socio-economic well-being

C. Depleting natural resources rapidly

D. Increasing dependency on chemical inputs

Answer: B. Balancing ecological health and socio-economic well-being

Chapter 10

Design and Diagnostic Study of Agroforestry System

The Design and Diagnostic (D&D) study of an agroforestry system is a systematic and participatory approach used to understand existing land use practices and to develop suitable agroforestry interventions tailored to local conditions. The process begins with a thorough diagnostic phase, where researchers, extension agents, and farmers collaborate to analyze the physical, biological, and socio-economic aspects of the current farming systems. This includes identifying constraints such as low soil fertility, water scarcity, poor productivity, and lack of income diversification. The diagnostic study helps in recognizing the key problems and potential opportunities for improvement through agroforestry. Once the problems are clearly understood, the design phase follows, focusing on the development of site-specific agroforestry systems that are technically feasible, socially acceptable, and economically viable. This includes selecting appropriate tree and crop species, determining their spatial and temporal arrangements, and planning suitable management practices. The D&D approach emphasizes flexibility, allowing for continuous refinement based on farmer feedback and field observations. It also identifies research gaps and extension needs to ensure long-term sustainability and scalability. Overall, the D&D methodology serves as a foundation for implementing successful agroforestry systems that are integrated, multifunctional, and responsive to local needs and environmental conditions.

Unveiling the Potential: Designing and Diagnosing Agroforestry Systems for Success

Agroforestry, the integration of trees and shrubs into agricultural landscapes, holds immense potential for enhancing livelihoods, improving soil health, and mitigating climate change. However, simply planting trees alongside crops isn't a guaranteed recipe for success. A well-designed and carefully monitored agroforestry system is crucial to realizing its full benefits. This is where design and diagnostic studies become indispensable tools.

Why Design and Diagnostics Matter

Imagine building a house without a blueprint or a doctor prescribing medicine without a diagnosis. The result could be disastrous. Similarly, implementing an agroforestry system without a clear understanding of the local context and potential challenges can lead to failure. Design and diagnostic studies provide the framework for:

- ☆ **Understanding the Landscape:** These studies delve deep into the local ecological, economic, and social context. They identify existing farming practices, soil conditions, water resources, and market opportunities.
- ☆ **Identifying Needs and Opportunities:** By understanding the farmer's goals, resource constraints, and knowledge gaps, the study uncovers opportunities to integrate trees in a way that addresses specific needs. This might involve improving soil fertility, diversifying income sources, or providing shade for livestock.
- ☆ **Tailoring Agroforestry Systems:** Based on the gathered information, the study helps design agroforestry systems that are specifically tailored to the local conditions and farmer's objectives. This includes selecting appropriate tree and crop species, determining optimal planting arrangements, and developing management strategies.
- ☆ **Monitoring and Evaluation:** Designing for success also includes establishing monitoring protocols to track the system's progress. This allows for early identification of problems and adaptive management strategies to ensure the system remains productive and beneficial over time.

The Design and Diagnostic Process

A typical design and diagnostic study for agroforestry typically involves several key phases:

1. Context Assessment

- ☆ **Participatory Rural Appraisal (PRA):** Engaging with local communities through workshops, focus group discussions, and interviews to understand their needs, priorities, and traditional farming practices.
- ☆ **Resource Inventory:** Mapping existing land use patterns, soil types, water resources, and vegetation cover.
- ☆ **Market Analysis:** Identifying potential markets for agroforestry products and assessing existing value chains.

2. System Design

- **Species Selection:** Choosing tree and crop species that are compatible with each other and well-suited to the local environment. This includes considering factors like growth rates, nutrient requirements, and market value.
- **Spatial Arrangement:** Determining the optimal spacing and arrangement of trees and crops to maximize resource utilization and minimize competition.
- **Management Practices:** Developing recommendations for pruning, fertilization, pest control, and other management practices to ensure the system's long-term productivity.

3. Implementation and Monitoring

- **Farmer Training:** Providing farmers with the knowledge and skills they need to manage the agroforestry system effectively.
- **Establishment of Monitoring Plots:** Setting up experimental plots to track the growth and performance of trees and crops under different management practices.
- **Data Collection and Analysis:** Regularly collecting data on soil fertility, crop yields, tree growth, and economic performance.

4. Evaluation and Adaptation

- **Analyzing Monitoring Data:** Assessing the system's performance and identifying any problems or areas for improvement.
- **Adapting Management Practices:** Making adjustments to the system's design and management based on the monitoring data.
- **Disseminating Findings:** Sharing the study's findings with other farmers and researchers to promote the adoption of successful agroforestry practices.

Benefits of a Well-Designed Agroforestry System

By investing in a well-designed and diagnosed agroforestry system, farmers and communities can reap a multitude of benefits, including:

- **Increased Crop Yields:** Trees can improve soil fertility, provide shade, and reduce soil erosion, leading to higher crop yields.
- **Diversified Income:** Agroforestry systems can provide multiple income streams from timber, fruits, nuts, and other tree products.
- **Improved Soil Health:** Trees can help improve soil structure, increase organic matter, and enhance water infiltration.

- **Enhanced Biodiversity:** Agroforestry systems can create habitats for a variety of plants and animals, promoting biodiversity conservation.
- **Climate Change Mitigation:** Trees can sequester carbon dioxide from the atmosphere, helping to mitigate climate change.

Design and diagnostic studies are not just academic exercises; they are the foundation for building successful and sustainable agroforestry systems. By understanding the local context, identifying needs and opportunities, and tailoring systems to specific conditions, we can unlock the full potential of agroforestry to improve livelihoods, protect the environment, and build a more resilient future. As agroforestry continues to gain prominence as a key strategy for sustainable agriculture and rural development, the importance of comprehensive design and diagnostic studies cannot be overstated. Investing in these studies is an investment in the long-term success and sustainability of agroforestry systems worldwide.

Planning for Agroforestry – Constraints, Diagnosis and Design Methodology

Agroforestry Planning of Farms

Agroforestry is the management of interactions between trees, crops, and livestock on farm plots, with the primary aim of achieving the objectives set by the farmer or farming household. These interactions refer to the influence one component has on another. Not all interactions are beneficial—some may be negative and result in competition for vital resources. For instance, trees with dense canopies may cast shade on understory crops, reducing their growth and productivity, thereby creating a negative interaction.

The role of the farm manager is to harness positive interactions while minimizing or eliminating the negative ones. Effective agroforestry planning enables the manager to regulate these interactions to optimize production, enhance overall system value, and promote environmental conservation. This planning is applicable to farms of all sizes.

In developing countries, researchers are increasingly reaching out to farmers through on-farm experimentation. However, these partnerships remain somewhat one-sided, as scientists often collect information from farmers without establishing reciprocal channels for knowledge exchange. There is a need for a two-way communication system where insights about technology, research needs, and local priorities flow seamlessly in both directions.

Ultimately, it is the farmer or land user who decides whether to adopt an agroforestry technology based on its relevance and practicality in their specific context. To support agroforestry researchers, ICRAF (International Centre for Research in Agroforestry) has assembled multidisciplinary teams—including anthropologists, economists, and agricultural and forestry scientists—who work alongside male and female farmers to develop a systematic methodology for diagnosing land management problems and designing appropriate agroforestry interventions. This approach follows the fundamental logic that, just as in medicine, **diagnosis must precede treatment**.

Constraints of Agroforestry

Despite its benefits, agroforestry systems face several constraints:

- ☆ The presence of trees may reduce crop yields compared to monocropping systems.
- ☆ Tree canopies may block sunlight, leading to competition for light.
- ☆ Felling trees can damage nearby arable crops.
- ☆ Trees and crops may compete for moisture, especially if trees lack deep taproots.
- ☆ Some tree species act as hosts to pests that harm main crops.
- ☆ Agroforestry requires more intensive and skilled management.
- ☆ Longer gestation periods of trees delay returns to the farmer.
- ☆ Farmers often prioritize field crops over tree crops.
- ☆ Certain tree species release chemical exudates that inhibit the growth of nearby agricultural crops.

The Genesis of D&D

Agroforestry has been described as "a new name for an old practice." The **Diagnosis and Design (D&D)** methodology is an adaptation of pre-existing methods to address the unique demands and characteristics of agroforestry systems.

Several methodologies have been developed for holistic evaluation of land use systems, notably:

- ☆ **Farming Systems Research/Extension (FSR/E)**
- ☆ **Land Evaluation methodology**

FSR/E was created to address the shortcomings of top-down, technology-transfer approaches used during the Green Revolution. It emphasizes interdisciplinary collaboration and active farmer participation. In contrast, the D&D methodology, developed by ICRAF under the guidance of J.B. Raintree, specifically addresses agroforestry constraints and opportunities that are often overlooked in other methodologies. For example, in many FSR/E programs, trees within the farming system are often ignored or considered invisible.

What is D&D (Diagnosis and Design)?

D&D is a structured methodology developed by ICRAF for diagnosing land management issues and designing agroforestry solutions. It assists researchers and field workers in planning and implementing effective agroforestry research and development projects.

Key Features of the D&D Methodology

A. Flexibility

D&D is a flexible, adaptive discovery process that can be customized to meet the needs and resources of different users.

B. Speed

D&D can be applied rapidly during project planning through a quick appraisal and can be followed by in-depth analysis during implementation.

C. Repetition

D&D is an iterative and open-ended process. Since initial designs often require refinement, the process continues until satisfactory improvements are achieved.

Criteria of a Good Agroforestry Design

There is no substitute for a well-conceived design. A good agroforestry design should fulfill the following criteria:

A. Productivity

Agroforestry can enhance productivity in multiple ways—by increasing the output of tree products, improving crop yields, reducing input needs, increasing labor efficiency, diversifying production, and contributing to economic and biological efficiency.

B. Sustainability

Agroforestry seeks to improve the sustainability of land use systems. It meets conservation goals while also appealing to low-income farmers who may be less motivated by conservation itself.

C. Adaptability

Even the most technically advanced and environmentally sound agroforestry design is ineffective if it is not adopted by users. Thus, the design must align with both the **social** and **environmental** contexts of the target farming system.

Who Can Use the D&D Methodology?

- ☆ **Researchers**
- ☆ **Extension officers**
- ☆ **Government field workers**
- ☆ **Non-Governmental Organizations (NGOs)**

Basic Procedure of D&D

The D&D process follows a logical sequence, which can be broken into small, manageable steps. While these steps can be selectively applied depending on the context, the underlying hierarchical logic is robust and adaptable to most agroforestry design challenges.

Whenever the process seems too detailed or complex, users are encouraged to return to the core procedural framework as a guide to reorient themselves and understand their position in the overall methodology.

Table 10.1: Basic Procedure of D&D

D&D Stages	*Basic Questions to Answer*	*Key Factors to Consider*	*Mode of Inquiry*
Pre-diagnostic	Definition of the land use systems and site selection (which system to focus on?)	Distinctive combinations of resources, technology and land user objectives	Seeing and comparing The different land use systems
	How does the system work? (How is it organized, how does it function to achieve its objectives?)	Production objectives and strategies, arrangement of components	Analyzing and describing the system
Diagnostic	How well does the system work? (What are its problems, limiting constraints, problem- generating syndromes and intervention points?)	Problems in meeting system Objectives (production short-falls, sustainability problems Casual factors, constraints and interventions points	Diagnostic interviews and direct field observations Troubleshooting the problems, subsystems
Design and Evaluation	How to improve the system? (What is needed to improve system performance?)	Specifications for problem solving or performance enhancing interventions	Iterative design and evaluation of alternatives
Planning	What to do to develop and disseminate the improved system?	Research and development needs, extension needs	Research design project planning
Implemen-tation	How to adjust to new information?	Feedback from on-station research, on-farm trials and special studies	Re-diagnosis and re-design in the light of new information

D&D is a Niterative Process

The basic D&D process is repeated throughout the project implementation stage to refine the original diagnosis and improve the technology design in the light of new information from on-farm research trials, more rigidly controlled on station investigations, and eventual extension trials in a wider range of sites. As shown in the following flowchart, the iterative D&D process provides a basis for close feedback complementarily between different project components. By adjusting the plan of action to new information, the D&D process becomes self corrective.

In an integrated agroforestry research and extension programme, key decisions are made during periodic meetings where new results are assessed and the action plan is revised accordingly. This iterative process continues until the system design is sufficiently optimized and no further refinements are necessary. While stakeholders can join the cycle at any stage, the final fine-tuning and widespread adoption of the technology are most likely to be carried out by the farmers themselves.

Key Concepts of D&D

A) **D&D is system specific**

B) **Definition of the system for D&D purposes:** A land use system is defined as a distinctive combination of three interrelated factors: the landresources, exploited by a particular technology, to satisfy the production objectives of a particular type of land user.

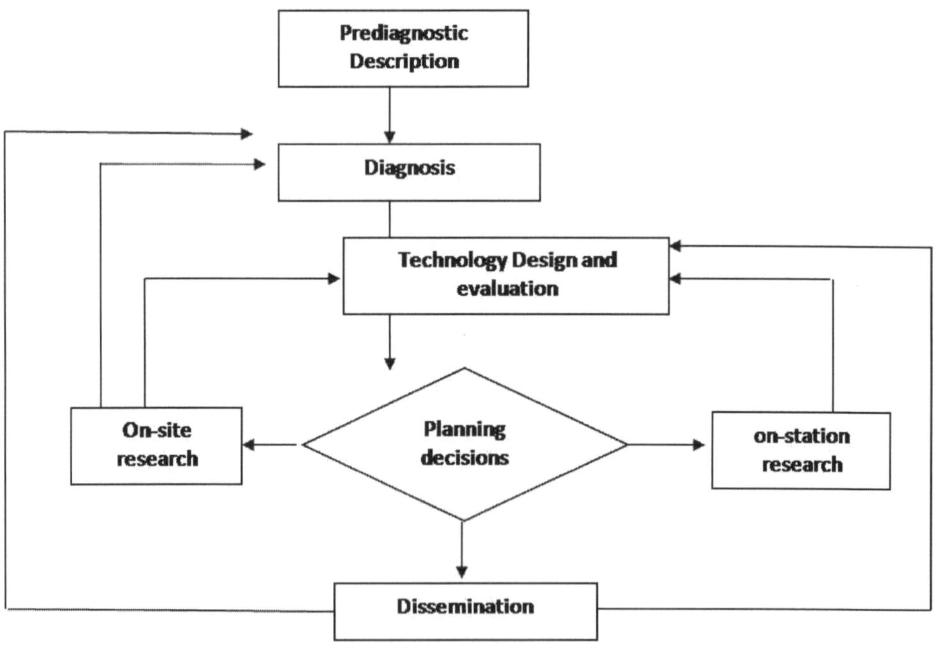

Figure 10.1: D&D is a Niterative Process.

C) **Diagnosis Leads to Specifications for Interventions:** The final output of the diagnostic phase is a set of **functional specifications**. These define:

- **What the system needs** – Identifying the key constraints, gaps, and opportunities.
- **How these needs can be met** – Outlining the requirements for interventions to improve performance, sustainability, and adoptability.

D) **Specifications Lead to Candidate Technologies:** Once the system's needs and constraints are specified, suitable **candidate technologies** are identified. These are the various combinations of practices, species, and arrangements that could meet the functional specifications of the system.

E) **Technology Specifications Complete the Design:** At this stage, the **detailed design** of the technology is completed. This includes:

- Selection of **component species** (plants and animals).
- **Spatial arrangements** of components within the system (farm layout, landscape zoning, *etc.*).
- **Management practices** necessary to ensure effectiveness, sustainability, and adoption.

F) **Design Reveals Research Needs and Extension Opportunities:** The design process often identifies **knowledge gaps**, which highlight the need for:

- **Further research** to refine technology or address uncertainties.
- **Extension strategies** to support adoption, including farmer training, demonstration plots, and communication of benefits.

G) **If at First You Don't Succeed, Try and Try Again:** Agroforestry system design is inherently **iterative**. Initial designs are often incomplete or imperfect. Continuous feedback from the field, adaptive experimentation, and participatory processes are essential to refine and improve both diagnosis and design.

Planning for Agroforestry: Constraints, Diagnosis and Design Methodology

I. Basic Principles and Procedures

Step 1: Planning the Study

- Define **objectives** of the study.
- Specify the **geographic area** to be covered.
- Identify **collaborating institutions** and team members.

- ☆ Select and adapt appropriate **Diagnosis and Design (D&D)** methodologies.

Step 2: Regional Reconnaissance

- ☆ Map and describe **major land units**.
- ☆ Analyze **population distribution** and landscape features.

Step 3: Identification and Preliminary Description of Land Use Systems

- ☆ Identify and characterize **key land use systems**.
- ☆ Assess their **constraints** and **problems**.
- ☆ Evaluate their **agroforestry potential**.

Step 4: Site Selection

Select priority land use systems based on:

- ☆ Severity of problems.
- ☆ Potential for agroforestry interventions.
- ☆ Regional representativeness.
- ☆ Choose **representative sites** for detailed D&D work.

II. Diagnostic Stage

Step 5: Diagnostic Survey

- ☆ Conduct field surveys of selected management units.
- ☆ Identify current **land use strategies** and **production problems**.
- ☆ Investigate **interactions** between ecological and social systems.

Step 6: Diagnostic Analysis

- ☆ Analyze field data to:
  - ❒ Identify **key constraints** and **development entry points**.
  - ❒ Assess **sustainability concerns**.

Step 7: Specification of Appropriate Interventions

- ☆ **Functional specifications** for interventions.
- ☆ **Design constraints** and limitations.
- ☆ **Desired attributes** of new technologies.
- ☆ An **overall development strategy** for the land use system.

III. Technology Design Stage

Step 8: Identification of Candidate Technologies

- ☆ List technologies that **meet system specifications**.
- ☆ Select and prioritize the most **promising combinations**.

Step 9: Detailed Technology Specifications

- ☆ Describe the **desired attributes** of each technology:
  - ❒ Component species and traits.
  - ❒ Required management practices.
- ☆ Prioritize these based on system needs and diagnostic findings.

Step 10: Technology Design

For each selected technology:

- ☆ Define:
  - ❒ **Functions** to be addressed.
  - ❒ **Placement** within the landscape or farm.
  - ❒ Best **components** to use.
  - ❒ **Quantity** of each component.
  - ❒ Specific **spatial arrangements**.
  - ❒ Required **management practices**.
- ☆ Identify unresolved questions for **further research**.
- ☆ Integrate findings into a **comprehensive agroforestry design** (consider stepwise adoption by farmers).

IV. Evaluation and Redesign Stage

Step 11: Ex-Ante Evaluation and Redesign

- ☆ Evaluate **user response** (optional verification survey).
- ☆ Compare agroforestry design against alternatives using:
  - ❒ **Productivity** (yield, economics, diversity).
  - ❒ **Sustainability** (environmental and resource impact).
  - ❒ **Adoptability** (cultural fit, benefits distribution).
- ☆ Redesign based on evaluation feedback.

Step 12: Suitability Classification

- ☆ Summarize evaluations and classify designs for **broader application**.
- ☆ Create **suitability maps and tables** to define recommendation domains.

V. Planning Stage

Step 13: Knowledge Review and Research Needs Assessment

- ☆ Assess readiness for **extension** or need for more research.
- ☆ List research needs, including:
  - ❒ Further D&D or follow-up.
  - ❒ On-farm trials and monitoring.
  - ❒ **Farmer-managed** and **researcher-managed** trials.

Step 14: Research and Extension Plan

- ☆ Develop an **action plan** including:
  - ❒ Specific **research activities**.
  - ❒ Planned **extension efforts**.
  - ❒ **Integration** of research and extension.
  - ❒ **Collaboration** within networks.

VI. Implementation Stage

Step 15: Implementation of R and D and Extension

- ☆ Use iterative D&D to **refine prototypes** based on feedback.
- ☆ Institutionalize **communication** among research, extension, and field teams.
- ☆ Hold **regular meetings** for learning and adjusting plans.

Comparison of D&D with Similar Methodologies

D&D vs. FSR/E (Farming Systems Research/Extension)

- ☆ **D&D** focuses more specifically on the role of **trees** in land use systems.
- ☆ More detailed and **elaborate design steps** for complex agroforestry landscapes.
- ☆ **Flexible** in scale of application.
- ☆ Strong emphasis on the **iterative nature** of diagnosis and design.

D&D vs. Land Evaluation

- ☆ **Land Evaluation** emphasizes **environmental factors**.
- ☆ **D&D** gives greater weight to **social dynamics** and participatory design.

D&D vs. Agro-ecosystem Analysis

- ☆ Agro-ecosystem analysis is simpler, good for **rapid rural appraisals**.
- ☆ Both share a **holistic and systems-based** approach.

General Observations

- All methodologies aim to **improve land use** and support decision-making.
- Each has **specific strengths** suited to different contexts.
- **D&D's agroforestry focus** makes it widely used in this domain.
- Success depends on:
  - Quality of **interdisciplinary teams**.
  - Strength of **available technologies**.
  - Effective **translation of diagnosis into action**.

QUESTION BANK

1. **What is the main purpose of conducting a diagnostic study in an agroforestry system?**

   A. To increase chemical fertilizer use

   B. To identify constraints and opportunities for system improvement

   C. To convert agroforestry to monoculture farming

   D. To reduce tree density

Answer: B. To identify constraints and opportunities for system improvement

2. **Which of the following is a key component in designing an agroforestry system?**

   A. Selecting appropriate tree and crop species based on site conditions

   B. Ignoring soil fertility status

   C. Removing livestock from the system

   D. Using only exotic species

Answer: A. Selecting appropriate tree and crop species based on site conditions

3. **Agroforestry system design should primarily consider:**

   A. Only economic returns

   B. Soil type, climate, and socio-economic conditions

   C. Maximum tree planting without crop integration

   D. Urban land use

Answer: B. Soil type, climate, and socio-economic conditions

4. **The diagnostic study in agroforestry includes:**

   A. Soil analysis, socio-economic survey, and vegetation inventory

   B. Only harvesting timber

   C. Selling land for urban development

   D. Ignoring farmer preferences

Answer: A. Soil analysis, socio-economic survey, and vegetation inventory

5. **Which design principle helps to minimize competition between trees and crops in agroforestry?**

   A. Planting trees in dense clusters

   B. Spatial and temporal arrangement of species

C. Using only annual crops

D. Eliminating crop rotation

Answer: B. Spatial and temporal arrangement of species

6. **Silvicultural practices in agroforestry design focus on:**

A. Enhancing growth and productivity of trees and crops

B. Removing all undergrowth

C. Planting trees without maintenance

D. Avoiding pruning or thinning

Answer: A. Enhancing growth and productivity of trees and crops

7. **What is an important factor to diagnose before implementing an agroforestry system?**

A. Market demand for tree and crop products

B. Availability of high-cost machinery only

C. Ignoring local climatic conditions

D. Monoculture dominance

Answer: A. Market demand for tree and crop products

8. **The key outcome of a well-conducted diagnostic study is:**

A. Development of site-specific agroforestry models

B. Random planting of trees and crops

C. Exclusion of local communities

D. Land degradation

Answer: A. Development of site-specific agroforestry models

9. **Which tool is often used in the diagnostic study of agroforestry systems?**

A. SWOT analysis (Strengths, Weaknesses, Opportunities, Threats)

B. Guesswork only

C. Ignoring farmer input

D. Using outdated species lists only

Answer: A. SWOT analysis (Strengths, Weaknesses, Opportunities, Threats)

10. **In agroforestry design, "temporal arrangement" means:**

A. Planting trees and crops at the same time only

B. Scheduling planting and harvesting to reduce competition and optimize resource use

C. Planting only trees and no crops

D. Harvesting all crops at once

Answer: B. Scheduling planting and harvesting to reduce competition and optimize resource use

Chapter 11

Silviculture: Definition and Scope

Silviculture is defined as the art and science of controlling the establishment, growth, composition, health, and quality of forests to meet diverse needs and values. It involves the practice of growing and managing forest trees systematically to achieve specific objectives, whether ecological, economic, or social. The primary aim of silviculture is to ensure the continuous production of forest goods and services such as timber, fuelwood, fodder, and non-timber forest products, while also conserving biodiversity and maintaining ecological balance.

The scope of silviculture is broad and multidisciplinary. It includes various activities such as regeneration (both natural and artificial), thinning, pruning, harvesting, and tending operations that influence the structure and productivity of forest stands. Silviculture also involves the selection of suitable tree species based on site conditions, understanding tree-soil-climate relationships, and designing appropriate silvicultural systems like clear felling, shelterwood, and selection systems.

In agroforestry systems, silviculture plays a crucial role by helping integrate tree species that complement crops and livestock, ensuring sustainable land use and improving farm income. It also contributes to soil conservation, carbon sequestration, climate change mitigation, and landscape restoration. With growing concerns over deforestation, land degradation, and biodiversity loss, the importance of silviculture has increased as a tool for sustainable forest and land management. Ultimately, silviculture serves as a foundational discipline within forestry and agroforestry, bridging ecological principles with practical management for long-term productivity and sustainability.

Silviculture: Shaping the Forest for Desired Outcomes

Silviculture, quite simply, is the art and science of controlling the establishment, growth, composition, health, and quality of forests and woodlands to meet the diverse needs and values of landowners and society. It's about actively managing forests, not just letting them grow wild, to achieve specific objectives, whether they be timber production, wildlife habitat enhancement, watershed protection, or recreational opportunities.

Understanding the definition is just the starting point. Delving into the **scope of silviculture** reveals a complex and dynamic field that blends ecological principles, practical application, and long-term planning. Here's a breakdown of the key areas within the silvicultural realm:

1. Forest Regeneration

One of the cornerstones of silviculture is ensuring the continuity of the forest. This involves implementing strategies to establish new trees, either naturally or artificially.

- **Natural Regeneration:** This relies on the dispersal of seeds from existing trees and the subsequent germination and establishment of seedlings. Silvicultural techniques can manipulate the environment to favor natural regeneration, such as creating seedbeds by removing competing vegetation or using prescribed burns.
- **Artificial Regeneration:** This involves actively planting seedlings or directly seeding areas. It's often employed when natural regeneration is unlikely to succeed, such as after a clearcut or in areas with limited seed sources. Selecting the right species for planting based on site conditions and desired outcomes is crucial.

2. Stand Dynamics and Manipulation

Once trees are established, silviculture focuses on influencing their growth and development. This involves understanding the interactions between trees within a stand and the impact of various silvicultural treatments.

- **Thinning:** This involves removing some trees from a stand to reduce competition for resources like sunlight, water, and nutrients. Thinning can promote the growth of remaining trees, improve their health, and increase the overall value of the stand.
- **Pruning:** Removing lower branches from trees can improve the quality and value of timber by reducing the size of knots.
- **Fertilization:** In nutrient-poor sites, fertilization can be used to boost tree growth and productivity.

3. Species Composition Management

Silviculture allows us to influence the types of trees that make up a forest. This is critical for achieving specific objectives.

- **Species Selection:** Choosing the right species for a particular site is fundamental to success. Factors to consider include the soil type, climate, and desired products or services.
- **Weeding:** Removing unwanted vegetation, including competing tree species, can favor the growth of the desired species.
- **Conversion:** Silvicultural treatments can be used to gradually convert a forest from one species composition to another.

4. Forest Health and Protection

Maintaining the health and resilience of the forest is a crucial aspect of silviculture.

- **Pest and Disease Management:** Implementing strategies to prevent or control outbreaks of forest pests and diseases is essential for protecting the forest from damage.
- **Fire Management:** Silvicultural practices can be used to reduce the risk of wildfires and to manage forests after a fire.

5. Harvesting and Utilization

While often viewed as separate, the harvesting of timber is an integral part of the silvicultural process.

- **Harvesting Methods:** Different harvesting methods, such as clearcutting, shelterwood cutting, and selection cutting, have different impacts on the forest ecosystem. Choosing the appropriate method is critical for achieving silvicultural objectives.
- **Utilization of Forest Products:** Ensuring efficient utilization of harvested wood and other forest products contributes to the sustainability of the forest.

6. Monitoring and Evaluation

Silviculture is not a static process. Regular monitoring and evaluation are essential to assess the effectiveness of silvicultural treatments and to adapt management strategies as needed.

Why is Silviculture Important?

The scope of silviculture extends far beyond simply growing trees for timber. It's about:

- ☆ **Sustainable Resource Management:** Ensuring that forests can provide a continuous supply of timber, water, wildlife habitat, and other resources for future generations.
- ☆ **Maintaining Biodiversity:** Managing forests to protect and enhance the diversity of plant and animal life.
- ☆ **Climate Change Mitigation:** Enhancing carbon sequestration in forests to help mitigate climate change.
- ☆ **Economic Benefits:** Providing economic opportunities for landowners and communities through timber production and other forest-related activities.
- ☆ **Ecological Restoration:** Using silvicultural techniques to restore degraded forests and ecosystems.

In conclusion, silviculture is a vital discipline that plays a crucial role in shaping the future of our forests. By understanding the principles and practices of silviculture, we can manage forests in a sustainable and responsible manner, ensuring that they continue to provide a wide range of benefits for society. The field is constantly evolving as we learn more about forest ecosystems and develop new technologies and techniques, making it a dynamic and crucial area of study.

QUESTION BANK

1. **Which of the following best defines silviculture?**
   A. The study of insects in forest ecosystems
   B. The art and science of cultivating forest trees
   C. The process of animal husbandry in forests
   D. The method of soil testing in forest lands

Answer: B. The art and science of cultivating forest trees

2. **The primary aim of silviculture is to:**
   A. Improve rainfall patterns
   B. Manage forest resources for sustained production
   C. Enhance fish production in forest rivers
   D. Eliminate natural forest species

Answer: B. Manage forest resources for sustained production

3. **Which of the following is NOT included in the scope of silviculture?**
   A. Tree planting
   B. Harvesting of trees
   C. Livestock breeding
   D. Thinning and pruning

Answer: C. Livestock breeding

4. **Silviculture contributes to agroforestry by:**
   A. Promoting monoculture systems
   B. Facilitating integration of trees with crops and livestock
   C. Replacing traditional farming practices
   D. Eliminating biodiversity from farmlands

Answer: B. Facilitating integration of trees with crops and livestock

5. **Which operation is directly associated with silviculture?**
   A. Fertilizer application in rice fields
   B. Pruning of forest trees
   C. Pollination of fruit crops
   D. Irrigation of sugarcane

Answer: B. Pruning of forest trees

6. **Silviculture differs from general forestry in that it:**
   A. Excludes natural forests
   B. Focuses on tree cultivation and stand management

C. Is concerned only with wildlife management

D. Is limited to dryland agriculture

Answer: B. Focuses on tree cultivation and stand management

7. **Which of the following systems is a part of silvicultural practices?**

A. Relay cropping system

B. Shelterwood system

C. Hydroponic system

D. Dairy management system

Answer: B. Shelterwood system

8. **Silviculture helps in achieving ecological goals such as:**

A. Increasing dependence on synthetic fertilizers

B. Promoting soil erosion

C. Enhancing biodiversity and ecological balance

D. Reducing forest canopy cover

Answer: C. Enhancing biodiversity and ecological balance

9. **Which of the following is a silvicultural objective in an agroforestry context?**

A. To eliminate all tree-crop interactions

B. To grow trees that compete with agricultural crops

C. To select and manage tree species that support crop and livestock productivity

D. To ensure only commercial timber production

Answer: C. To select and manage tree species that support crop and livestock productivity

10. **In the context of silviculture, the term 'regeneration' refers to:**

A. The destruction of forests for agriculture

B. The natural or artificial renewal of forest stands

C. The fertilization of agricultural fields

D. The expansion of animal rearing zones

Answer: B. The natural or artificial renewal of forest stands

Chapter 12

Propagation of Tree Species, Management of Tree Species

From Seed to Stand: Propagation and Management of Tree Species

Trees are essential components of our planet, providing countless benefits from clean air and water to habitat and carbon sequestration. Understanding how to propagate and manage tree species is crucial for maintaining healthy forests, restoring degraded landscapes, and ensuring a sustainable future. This article explores the fundamentals of tree propagation and highlights key management practices.

Propagation: Starting Life Afresh

Tree propagation is the process of creating new trees from existing ones. This can be achieved through various methods, broadly categorized into sexual and asexual propagation:

1. Sexual Propagation (Seed Propagation)

This is the most common and natural method, involving the use of seeds.

- **Seed Collection:** The key to successful propagation lies in collecting viable seeds from healthy, mature trees. Timing is crucial, as seeds must be collected when they are ripe.
- **Seed Processing:** Once collected, seeds often require cleaning to remove pulp and extraction from cones. Drying and storage are essential for preserving viability. Some species require specific treatments to break dormancy, such as stratification (cold, moist storage) or scarification (weakening the seed coat).

- **Seed Sowing:** Seeds can be sown directly into the ground (direct seeding) or in nurseries for later transplanting. Nursery seedlings offer better control over growing conditions and can result in higher survival rates.
- **Factors Affecting Success:** Soil quality, moisture, sunlight, and temperature all play crucial roles in seed germination and seedling establishment. Protection from pests and diseases is also vital.

2. Asexual Propagation (Vegetative Propagation)

This method involves creating new trees from vegetative parts of the parent plant, ensuring genetic uniformity.

- **Cuttings:** Stem, root, or leaf cuttings are used to regenerate a new plant. Cuttings are treated with rooting hormones and placed in a suitable medium until roots develop.
- **Grafting:** A scion (a twig or bud) from a desirable tree is joined to the rootstock of another tree. Grafting is commonly used for fruit trees and ornamental trees to combine desirable traits like disease resistance and fruit quality.
- **Layering:** A branch of a parent tree is encouraged to root while still attached. Once roots develop, the rooted branch is separated and planted as a new tree.
- **Tissue Culture (Micropropagation):** This advanced technique involves growing plant tissues in a sterile environment. It allows for the rapid multiplication of selected trees and is particularly useful for rare or difficult-to-propagate species.

Management: Nurturing Growth and Sustainability

Once trees are established, ongoing management is essential for their health, productivity, and the overall integrity of the ecosystem. Tree management practices vary depending on the species, location, and desired outcomes.

- **Site Preparation:** Preparing the planting site is critical for seedling survival and growth. This may involve clearing vegetation, tilling the soil, and amending nutrient deficiencies.
- **Weed Control:** Weeds compete with seedlings for water, nutrients, and sunlight. Controlling weeds through manual removal, mulching, or herbicides is vital, especially during the early years.
- **Thinning:** Removing weaker or overcrowded trees allows the remaining trees to grow faster and healthier. Thinning also improves light penetration and reduces competition for resources.

- ☆ **Pruning:** Removing dead, diseased, or unwanted branches improves tree structure, encourages fruiting (in fruit trees), and enhances aesthetic appeal.
- ☆ **Fertilization:** Applying fertilizers can correct nutrient deficiencies and promote healthy growth. Soil testing should be conducted to determine the specific nutrient needs of the trees.
- ☆ **Pest and Disease Management:** Monitoring for pests and diseases and implementing control measures is crucial for preventing significant damage. Integrated Pest Management (IPM) strategies, which combine biological, cultural, and chemical controls, are preferred for minimizing environmental impact.
- ☆ **Water Management:** Ensuring adequate water availability is essential, especially during dry periods. Irrigation may be necessary in some regions.
- ☆ **Fire Management:** In fire-prone areas, implementing fire prevention and suppression measures is crucial for protecting trees and forests.
- ☆ **Harvesting (if applicable):** Sustainable harvesting practices ensure that timber is harvested in a way that maintains the long-term health and productivity of the forest. Reforestation after harvesting is essential for maintaining forest cover.

Understanding the principles of tree propagation and management is fundamental for ensuring the health and sustainability of our forests and landscapes. By employing appropriate propagation techniques and implementing sound management practices, we can contribute to the vital role trees play in our environment and our lives. This knowledge is essential for foresters, landowners, and anyone interested in the future of our planet. As we face increasing environmental challenges, the ability to propagate and manage tree species effectively will be more important than ever.

Selection of Tree Crops species for Agroforestry

Agroforestry is a deliberate integration of trees and crops in general, in same unit of land. These trees and crops compete with each other for nutrients, moisture and light. Therefore there are both +ve and –ve interaction among different components. In order to have a positive interaction among different components one must select a compatible component so that from a single unit of land a farmer/cultivator maximize his production. While selecting tree species for Agroforestry three factors must taken into consideration such as.

- ☆ Climate
- ☆ Soil
- ☆ Bioticfactor

A. Tree Species Selection Based on Climate Zones

In agroforestry systems, tree species selection must align with the climatic conditions of the region to ensure survival, productivity, and ecological compatibility. Below is a classification of recommended species based on climatic zones:

1. Hot Desert

- ✰ *Prosopis cineraria, P. chilensis, Acacia tortilis, Capparis spp., Tecomella undulata*

2. Cold Desert

- ✰ *Populus nigra, P. ciliata, P. alba, P. tremula, P. euphratica, Salix alba, S. fragilis, Juniperus spp.*

3. Tropical Semi-Arid

- ✰ *Prosopis spp., Acacia tortilis, A. nilotica, A. senegal, Albizia lebbeck, Eucalyptus camaldulensis, Azadirachta indica, Salvadora persica, Tamarix spp.*

4. Subtropical Semi-Arid

- ✰ *Pinus roxburghii, Acacia modesta, Albizia procera, Bauhinia variegata, Morus indica, Ficus spp.*

5. Temperate Semi-Arid

- ✰ *Pinus gerardiana, Juniperus macropoda, Corylus colurna*

6. Humid Tropical

- ✰ *Terminalia myriocarpa, Tectona grandis, Terminalia alata, Schima wallichii, Gmelina arborea, Dipterocarpus macrocarpus, Cocos nucifera, Areca catechu, Artocarpus heterophyllus, Pterocarpus santalinus, Chukrasia tabularis*

7. Humid Subtropical

- ✰ *Eucalyptus globulus, Acer oblongum, Acrocarpus fraxinifolius, Aesculus indica, Pinus kesiya, Prunus spp., Quercus spp.*

8. Humid Temperate

- ✰ *Acer campbellii, Abies pindrow, Quercus spp., Robinia pseudoacacia, Pinus wallichiana, Alnus nitida, Populus ciliata, Cryptomeria japonica*

9. Subtropical Semi-Humid

- ✰ *Albizia chinensis, Pinus roxburghii, P. kesiya, P. elliottii, Grewia optiva, Celtis australis, Eucalyptus grandis, E. globulus, Toona ciliata*

10. Tropical Sub-Humid

- ☆ *Eucalyptus tereticornis, E. citriodora, Casuarina equisetifolia, Dalbergia latifolia, Bombax ceiba, Morus alba, Leucaena leucocephala, Dalbergia sissoo, Anthocephalus chinensis, Adina cordifolia, Populus deltoides, Moringa oleifera*

11. Temperate Semi-Humid

- ☆ *Acacia mearnsii, Acer oblongum, Alnus nepalensis, Cedrus deodara, Celtis australis, Fraxinus spp., Quercus spp., Juglans regia*

B. Tree Species Selection Based on Soil Type

1. Desert Soils

- ☆ *Prosopis cineraria, P. chilensis, Acacia tortilis, A. senegal, A. nilotica, Salvadora spp.*

2. Recent Alluvium

- ☆ *Acacia catechu, Dalbergia sissoo, Bombax ceiba*

3. Saline-Alkali Soils

- ☆ *Prosopis spp., Acacia nilotica, Azadirachta indica, Ailanthus spp., Eucalyptus spp., Tamarix spp., Pongamia pinnata*

4. Coastal and Deltaic Alluvium

- ☆ *Casuarina equisetifolia, Cocos nucifera, Areca catechu, Avicennia spp.*

5. Red Soils

- ☆ *Tectona grandis, Madhuca indica, Mangifera indica, Dalbergia sissoo, Acacia nilotica, Leucaena leucocephala, Azadirachta indica, Eucalyptus hybrid, Pterocarpus marsupium, Adina cordifolia, Dendrocalamus strictus*

6. Black Cotton Soils

- ☆ *Acacia nilotica, A. leucophloea, Tectona grandis, Hardwickia binata, Adina cordifolia, Tamarindus indica, Aegle marmelos, Bauhinia spp., Dalbergia latifolia*

7. Laterite and Lateritic Soils

- ☆ *Tectona grandis, Eucalyptus spp., Acacia auriculiformis, Azadirachta indica, Tamarindus indica, Emblica officinalis*

8. Peaty and Organic Soils

- ☆ *Syzygium cuminii, Ficus glomerata, Bischofia javanica, Lagerstroemia speciosa, Gliricidia sepium*

9. Hill Soils

- ☆ *Juglans regia, Alnus nitida, Toona serrata, Cedrus deodara, Quercus spp., Grewia optiva, Celtis australis*

C. Biotic Factors Affecting Tree Species Selection

Tree species selection in agroforestry is also influenced by biotic stresses such as:

- ☆ **Grazing pressure**
- ☆ **Fire hazards**
- ☆ **Incidence of pests and diseases**

Species chosen should be resistant or tolerant to such biotic factors to ensure sustainability and minimal maintenance.

D. Desirable Characteristics of Tree Species for Agroforestry

Tree species suitable for agroforestry should ideally possess the following characteristics:

1. Non-competitive for Soil Moisture

- ☆ Low water requirements
- ☆ Deep taproot systems to avoid competing with crops for surface moisture

2. Non-competitive for Soil Nutrients

- ☆ Low nutrient demand
- ☆ Contribution to soil fertility, especially through nitrogen fixation (*e.g.*, leguminous trees)

3. Minimal Competition for Light

- ☆ Light-branching habit
- ☆ Allow sufficient sunlight penetration for crops underneath
- ☆ Capability to tolerate pruning if canopy is dense

4. High Survival Rate and Easy Establishment

- ☆ High transplant survival percentage
- ☆ Quick root regeneration

- ☆ Low mortality rate
- ☆ Hardy nature to withstand transplant shock

5. **Fast Growth and Easy Management**
   - ☆ Rapid early growth
   - ☆ Short rotation periods
   - ☆ Easy to manage species like *Poplar, Casuarina, Leucaena leucocephala*

6. **Wide Adaptability**
   - ☆ Suitability across varied agro-climatic zones

7. **High Palatability for Fodder**
   - ☆ Palatable and digestible foliage for livestock
   - ☆ Suitable for cut-and-carry fodder systems

8. **Soil Stabilization and Shelter Attributes**
   - ☆ Effective in preventing erosion and stabilizing soil
   - ☆ Examples include *Populus spp., Salix spp., Casuarina equisetifolia*

9. **Tolerance to Management Practices**
   - ☆ Ability to withstand pruning, lopping, and pollarding without loss of vigor

10. **Nutrient Cycling and Nitrogen Fixation**
    - ☆ Ability to recycle deep-soil nutrients through leaf litter
    - ☆ Enhancement of topsoil fertility
    - ☆ Nitrogen-fixing species (*e.g., Leucaena, Prosopis, Acacia, Casuarina*) contribute to soil nitrogen status through root nodules in symbiosis with *Rhizobium* bacteria

The following are a few tree species which help in fixing atmospheric nitrogen through their roots.

Table 12.1: Nitrogen Fixing Tree Species

1.	*Acacia albida*	21.	*Bauhinia variegata*
2.	*Acacia auriculiformis*	22.	*Butea monosperma*
3.	*Acacia catechu*	23.	*Cassia fistula*
4.	*Acacia aneura*	24.	*Cassia siamea*
5.	*Acacia dealbata*	25.	*Casuarina equisetifolia*
6.	*Acacia decurrens*	26.	*Dalbergia latifolia*

7.	*Acacia farnesiana*	27.	*Dalbergia sissoo*
8.	*Acacia implexa*	28.	*Delonix regia*
9.	*Acacia leucophloea*	29.	*Gliricidia sepium*
10.	*Acacia mearnsii*	30.	*Hardwicki abinata*
11.	*Acacia melanoxylon*	31.	*Leucaena leucocephala*
12.	*Acacia mollissima*	32.	*Moringa oleifera*
13.	*Acacia nilotica*	33.	*Oogeinia oojeinensis*
14.	*Acacia planifrons*	34.	*Parkinsonia aculeata*
15.	*Acacia senegal*	35.	*Peltophorum ferrugineum*
16.	*Albizia chinensis*	36.	*Pithecell obiumdulce*
17.	*Albizia lebbek*	37.	*Prosopis alba*
18.	*Albizia procera*	38.	*Prosopis chilensis*
19.	*Alnus nepalensis*	39.	*Prosopis cineraria*
20.	*Alnus nitida*	40.	*Robinia pseudoacacia*
41.	*Samane asaman*	44.	*Sesbania bispinosa*
42.	*Saraca indica*	45.	*Sesbania grandiflora*
43.	*Sesbania aegyptica*	46.	*Tamarindus indica*

Desirable Characteristics of Tree Species for Agroforestry Systems

In agroforestry systems, the choice of tree species is crucial for successful integration with crops and livestock. The selected species should exhibit certain traits that enhance productivity, soil health, and compatibility with associated crops.

- ☆ **Thin Bark and Low Bark Shedding:** Tree species chosen for agroforestry should have thin bark and should not shed it frequently. Frequent bark shedding can lead to the accumulation of debris, creating unhygienic conditions and potentially hampering the growth of the understorey crops.
- ☆ **Absence of Chemical Exudates:** Ideal tree species for agroforestry should be free from chemical exudations, particularly allelochemicals. These compounds, when released into the soil, can suppress the germination and growth of adjacent agricultural crops, affecting overall productivity.
- ☆ **Leaves Should Be Easily Decomposable:** Trees with fast-decomposing leaves are more suitable for agroforestry systems. Leaves that break down quickly contribute organic matter and nutrients to the soil in a shorter time. Most leguminous trees have small, thin leaves that decompose easily, enriching the soil. In contrast, species with broad

and fibrous leaves, such as teak (*Tectona grandis*), mango (*Mangifera indica*), and banyan (*Ficus benghalensis*), decompose slowly. Moreover, when large leaves fall on tender crops, they can obstruct light penetration, hampering photosynthesis.

- **Multiple Uses of the Tree Species:** The selected tree species should offer multiple benefits. Trees that provide more than one useful product such as fuelwood, leaf fodder, edible fruits, flowers, timber, or fibre are more desirable. This multifunctionality ensures economic viability and resilience for the farmer.
- **High Yield Potential:** High productivity is a critical selection criterion for tree species in agroforestry. Since the goal is to maximize output per unit area, species with proven high-yield potential should be prioritized. Care must be taken to ensure that seeds or planting material are sourced from reliable and genetically superior sources.

Desirable Characteristics of Agricultural Crops for Agroforestry Systems

For successful integration in agroforestry, agricultural crops should also possess certain favorable traits:

- **Short Duration and Fast Growth:** Crops with a short growth cycle and quick maturity are preferred, as they can complete their life cycle without prolonged competition with trees.
- **Partial Shade Tolerance:** Since trees may cast shade, the crops should be able to tolerate at least partial shade to maintain good yield.
- **Belonging to the Leguminosae Family:** Crops from the legume family are particularly desirable because they fix atmospheric nitrogen and improve soil fertility.
- **Responsiveness to High-Density Tree Planting:** The crops should perform well even in systems with closely spaced tree plantings.
- **Tolerance to Environmental Stress:** Crops that can withstand mild drought or waterlogging are advantageous, especially in variable agro-climatic conditions.
- **Contribution of Organic Matter:** Ideal crops should contribute to soil health by returning organic matter through fallen leaves, roots, and stubble after harvest.
- **Compatibility with Multiple Cropping Systems:** Crops should be suitable for inclusion in intensive and diversified cropping systems, making efficient use of space and resources.

QUESTION BANK

Propagation of Tree Species

1. **Which of the following is a vegetative method of tree propagation?**
   A. Seed sowing B. Grafting
   C. Stratification D. Scarification

Answer: B. Grafting

2. **Which hormone is commonly used to promote root initiation in cuttings?**
   A. Cytokinin B. Ethylene
   C. Auxin (IBA/IAA) D. Gibberellin

Answer: C. Auxin (IBA/IAA)

3. **What is the primary benefit of vegetative propagation in trees?**
   A. Slower growth rate B. High genetic variability
   C. Uniformity in plant traits D. More seeds produced

Answer: C. Uniformity in plant traits

4. **Seed dormancy in many tree species can be broken by:**
   A. Stratification or scarification B. Sun drying
   C. Irrigation D. Fertilizer application

Answer: A. Stratification or scarification

5. **Which species is commonly propagated through coppicing?**
   A. Mango B. Teak
   C. Eucalyptus D. Tamarind

Answer: C. Eucalyptus

Management of Tree Species

6. **Pollarding is a tree management practice that involves:**
   A. Cutting the tree at ground level
   B. Removing only lateral branches
   C. Cutting upper branches to encourage regrowth
   D. Removing roots to limit growth

Answer: C. Cutting upper branches to encourage regrowth

7. **Thinning in agroforestry systems is done primarily to:**
   A. Increase competition among trees
   B. Reduce light for crops
   C. Minimize overcrowding and promote healthy growth
   D. Enhance seed dormancy

Answer: C. Minimize overcrowding and promote healthy growth

8. **Which silvicultural operation improves tree form and timber quality?**
   A. Weeding
   B. Pruning
   C. Topping
   D. Layering

Answer: B. Pruning

9. **In coppice management systems, new shoots regenerate from:**
   A. Seeds
   B. Branches
   C. Stump or root collar
   D. Leaf nodes

Answer: C. Stump or root collar

10. **The rotation period in tree management refers to:**
    A. The time required to sow seeds
    B. The age at which a tree starts flowering
    C. The period between planting and final harvest
    D. The interval between irrigation schedules

Answer: C. The period between planting and final harvest

Chapter 13

Regeneration by Seed, Coppice, Root Suckers, Transplanting, Stump, Branch Cutting, Rhizomes

Nature's Revival: Exploring the Diverse Methods of Plant Regeneration

The plant kingdom boasts an astonishing array of survival strategies, and chief among them is the ability to regenerate and propagate. From the humble seed to the complex network of rhizomes, plants have evolved diverse methods to ensure their survival and expansion. This article delves into some of the key techniques employed by plants to regenerate, highlighting the fascinating ingenuity of nature's design.

1. The Seed: A Miniature Time Capsule of Life

Perhaps the most well-known method, regeneration by seed is a fundamental process in the plant world. Seeds are essentially embryonic plants encased in a protective coating, packed with nutrients to fuel initial growth. Factors like wind, water, animals, and even explosive mechanisms aid in seed dispersal, allowing the plant to colonize new territories. Germination, the process of the seed sprouting and developing into a seedling, is triggered by favorable conditions such as moisture, temperature, and light.

2. Coppicing: Rebirth from the Stump

Coppicing is a remarkable form of vegetative regeneration where trees or shrubs are cut back to near ground level. In response, the remaining stump or stool sends up numerous shoots from dormant buds. This method is particularly common in deciduous trees like willow, hazel, and oak. Coppicing not only provides a sustainable source of timber and other materials but also promotes biodiversity by creating a mosaic of habitats.

3. Root Suckers: Expanding the Colony Underground

Some trees and shrubs can regenerate via root suckers. These are new shoots that arise from the roots of the parent plant. This method allows the plant to spread vegetatively, forming dense thickets or colonies. Aspen trees are a prime example of plants that rely heavily on root suckering, forming vast interconnected "forests" from a single original tree.

4. Transplanting: Moving Established Individuals

While not a form of natural regeneration in the strictest sense, transplanting involves moving an established plant from one location to another. This method is widely used in horticulture and agriculture. Careful excavation, minimal root disturbance, and appropriate post-transplant care are crucial for successful transplantation.

5. Stump Sprouting: Resilience in the Face of Adversity

Similar to coppicing, stump sprouting occurs when new shoots emerge from the remaining stump of a tree after it has been felled or damaged. This is a testament to the plant's resilience and internal reserves. Many hardwood trees, like maples and beeches, are capable of stump sprouting, allowing them to persist even after significant setbacks.

6. Branch Cuttings: Cloning Through Stem Segments

Branch cuttings are a common method of vegetative propagation, particularly in horticulture. A segment of a stem is cut from the parent plant and placed in a suitable rooting medium. With proper care, the cutting will develop roots and grow into a new, genetically identical plant. This method is widely used for propagating ornamental shrubs, fruit trees, and other desirable plant varieties.

7. Rhizomes: Underground Pioneers

Rhizomes are horizontal, underground stems that can produce new shoots and roots at intervals. They are a powerful adaptation for vegetative spread, allowing plants to colonize new areas rapidly. Plants like ginger, bamboo, and certain grasses rely heavily on rhizomes for regeneration and dispersal. This method allows them to thrive in challenging environments and quickly recover from disturbances.

The ability of plants to regenerate through diverse methods underscores their remarkable adaptability and resilience. From the intricate design of a seed to the sprawling network of rhizomes, each method represents a unique solution to the challenges of survival and propagation. Understanding these regenerative strategies not only deepens our appreciation for the plant kingdom but also provides valuable insights for conservation efforts, sustainable forestry, and even agricultural practices. By harnessing the power of nature's renewal, we can work towards a more sustainable and biodiverse future.

Different National and International Institutes Working in the Field of Agroforestry/Forestry

Introduction

In 1975, the International Development Research Centre (IDRC) of Canada commissioned a study by John Bene to examine forestry research and training worldwide. The study aimed to:

- ✰ Identify gaps in global forestry research and education;
- ✰ Evaluate the interconnection between agriculture and forestry in tropical, low-income countries;
- ✰ Propose research that enhances land use efficiency;
- ✰ Develop programs with economic and social benefits for developing nations;
- ✰ Recommend institutional structures for effective implementation;
- ✰ Formulate an action plan to attract support from international donors.

India

- ✰ Agroforestry research in India began in the 1960s–70s through institutes like **IGFRI, CSWCRTI, CRIDA, CAZRI**, and the **ICAR Complex for NE Region**.
- ✰ The **Seventh Five-Year Plan (1985–90)** recognized the importance of agroforestry and included it in the agricultural education curriculum.
- ✰ **Organized agroforestry research** was launched in the **Sixth Plan (1983)** through the **All India Coordinated Research Project on Agroforestry (AICRP-AF)** with 20 centres (12 at SAUs and 8 at ICAR institutes).
- ✰ The establishment of **NRCAF** was proposed during the Seventh Plan to enhance coordinated research.
- ✰ In the Seventh Plan, 11 more centres were added at SAUs.
- ✰ In the Eighth Plan, two new centres were included: **TANUVAS, Kattupakkam**, and **IGKV, Raipur**.
- ✰ Currently, **39 centres** (33 regular, 5 voluntary, and 1 PC unit) participate in coordinated agroforestry research across India.

International Institutions

1. Food and Agriculture Organization (FAO), Rome, Italy

- ✰ Coordinates global efforts on forestry genetic resource management.

- ☆ Promotes use and conservation of multipurpose species, especially in arid and semi-arid zones.
- ☆ Focuses on in situ and ex situ conservation and disseminates technical information via the *Forest Genetic Resources Information* newsletter.

2. International Board for Plant Genetic Resources (IBPGR), Rome, Italy

- ☆ A CGIAR institute focusing on conservation and improvement of plant genetic resources, including fuelwood species.

3. International Council for Research in Agroforestry (ICRAF), Nairobi, Kenya

- ☆ Established in 1977, now part of the World Agroforestry Centre.
- ☆ Aims to improve livelihoods in developing countries through agroforestry innovations.
- ☆ Focuses on research, training, and dissemination of agroforestry practices.

4. International Tropical Timber Organization (ITTO), Yokohama, Japan

- ☆ Promotes sustainable tropical timber trade and conservation.
- ☆ Supports projects on sustainable forest management and conservation of genetic resources.

5. International Union for Conservation of Nature (IUCN), Gland, Switzerland

- ☆ A unique body with both government and NGO members.
- ☆ Focuses on species conservation and genetic resource management.
- ☆ Publishes the *IUCN Red Data Book* and operates the *World Conservation Monitoring Centre (WCMC)* in the UK.

6. International Union of Forestry Research Organizations (IUFRO), Vienna, Austria

- ☆ Coordinates global forest research.
- ☆ Emphasizes industrial species improvement and forest genetic conservation.

7. United Nations Environment Programme (UNEP), Nairobi, Kenya

- ☆ Promotes environmental awareness and biodiversity conservation.
- ☆ Partners with governments and NGOs for global environmental monitoring.

8. United Nations Educational, Scientific and Cultural Organization (UNESCO), Paris, France

- Involved in gene conservation through the *Man and Biosphere (MAB)* Programme.
- Aims to conserve ecosystems and biological diversity.

9. Consultative Group on International Agricultural Research (CGIAR), Washington DC, USA

- Coordinates global agricultural research, supporting 13 international centres.
- Addresses conservation of plant and forest genetic resources.

National and Regional Institutions with International Programs

1. Central America and Mexico Coniferous Research Cooperative (CAMCORE)

- Promotes germplasm conservation of Central American and Mexican conifers and angiosperms.
- Distributes seeds for research, conservation stands, and progeny tests.

2. Centre Agronomic Tropical de Investigación y Enseñanza (CATIE), Costa Rica

- Works with Central and Latin American nations.
- Programs include agroforestry, silviculture, and watershed management.

3. Canadian International Development Agency (CIDA) and International Development Research Centre (IDRC), Canada

- Fund forestry research and development programs in developing nations.

4. Centre Technique Forestier Tropical (CTFT), France

- A branch of French foreign aid working in 12 African countries.
- Focuses on species such as Eucalyptus, Pinus, Terminalia, and Acacia.

5. Commonwealth Scientific and Industrial Research Organization (CSIRO), Canberra, Australia

- Tree Seed Centre supplies high-quality seeds of promising Australian woody plants for global research.

6. Danish Forest Seed Centre (DFSC), Humlebaek, Denmark

- ✰ Works in South-East Asia and Central America.
- ✰ Focuses on seed handling, storage, and conservation of arid-zone tree species.

7. Oxford Forestry Institute (OFI), UK

- ✰ Formerly known as the Commonwealth Forestry Institute.
- ✰ Supports international provenance testing and conservation of tropical species.

8. U.S. Department of Agriculture's Forest Service

- ✰ Conducts genetic improvement and conservation research in U.S. forests.
- ✰ Individual states run their own conservation programs for important tree species.

Notable Indian Institutions

9. National Research Centre for Agroforestry (NRCAF), Jhansi, Uttar Pradesh

- ✰ Conducts research and develops technologies for sustainable agroforestry.
- ✰ Coordinates with ICAR and SAUs.
- ✰ Provides training, consultancy, and information services.

10. Indian Grassland and Fodder Research Institute (IGFRI), Jhansi, Uttar Pradesh

- ✰ Conducts research in forage production and utilization.
- ✰ Supports sustainable agriculture and livestock productivity in South Asia.

Multipurpose Tree Species (MPTS) and their Management

Introduction to Multipurpose Tree Species (MPTS)

Multipurpose tree species (MPTS) refer to plant species that are intentionally cultivated to provide two or more distinct benefits or services. These may include tangible products such as fuelwood, food, fodder, fiber, timber, and medicinal compounds, as well as ecological services such as soil fertility enhancement, shade, shelter, microclimate regulation, and land sustainability. While many woody perennials may serve multiple functions in one system, their utility may be more limited in another, thus making context a key factor in their classification as MPTS.

Benefits Derived from MPTS

1. Food

- ☆ Trees provide a range of edible products for humans, such as fruits, nuts, edible leaves, and cereal substitutes.
- ☆ Tree leaves, pods, and fruits also serve as valuable livestock feed.
- ☆ Certain trees act as fertilizer trees, improving crop nutrition through nitrogen fixation, deep-root nutrient uptake, and enhanced soil organic matter.

2. Water Management

- ☆ MPTS contribute to soil moisture retention in rain-fed systems and pastures.
- ☆ They regulate stream flow, reduce flood risks, and improve water availability through better infiltration and runoff management.
- ☆ Trees act as living barriers to protect irrigation structures and help drain waterlogged or saline soils, especially phreatophytic species.
- ☆ Tree fodder, rich in moisture, supports livestock during dry periods.

3. Energy

- ☆ Trees serve as a renewable source of firewood and are used in pyrolytic conversion to produce charcoal, bio-oil, and gas.
- ☆ Wood and charcoal feedstocks are used in gasification processes, and carbohydrate-rich fruits can be fermented to produce ethanol.
- ☆ Tree-derived latex, oils, and resins contribute to bioenergy alternatives.
- ☆ Strategically placed trees can enhance wind energy efficiency by creating venturi effects.

4. Shelter and Construction

- ☆ Trees provide timber and poles for shelter and housing.
- ☆ They offer shade to humans, animals, and shade-tolerant crops.
- ☆ Windbreaks and shelterbelts protect agricultural fields, homesteads, and pastures.
- ☆ Living fences act as barriers and property boundaries.

5. Raw Materials

- ☆ MPTS supply raw materials for craft industries, fiber production, and food processing.
- ☆ They are sources of tannins, essential oils, dyes, medicinal compounds, and other industrial ingredients.

6. Economic Returns

- Farmers gain direct income from selling MPTS products and indirect gains through productivity enhancement and input cost savings.
- MPTS support livestock and crop productivity, leading to better financial returns from the integrated system.

7. Social and Cultural Value

- MPTS contribute to socio-cultural functions such as ceremonial use, gifting, and community-level trade.
- Income from MPTS can support social obligations like weddings, rituals, and local development.

Desirable Characteristics of MPTS

An ideal MPTS for agroforestry should have:

- Broad adaptability to local agro-climatic conditions.
- A thin, sparse crown that allows light penetration to understory crops.
- The ability to withstand management operations like coppicing, lopping, and pollarding.
- A quick sprouting habit post-pruning or harvesting.
- High productivity in terms of food, fodder, fuelwood, fiber, and medicinal value.
- Leaf litter that decomposes easily to return nutrients to the soil.
- Few or prunable lateral roots to reduce crop competition.
- Nitrogen-fixing capabilities (especially from leguminous species).
- Tolerance to environmental stresses such as drought, flooding, and soil variability.
- A deep taproot system for water extraction and soil stability.
- Easy establishment and low maintenance cost.
- High market demand and good value for produce.

Management Strategies for MPTS in Agroforestry Systems

System Design and Spatial Management

Once the appropriate combination of tree and agricultural components is selected, a management strategy that maximizes system performance must be implemented.

- **Horizontal management** (for water and nutrient-limited environments) includes:
  - Adjusting tree spacing.

- ❒ Designing zonal arrangements (macro/micro).
- ❒ Practicing mixed cropping patterns.

- ☆ **Vertical management** (for light-limited environments with adequate water and nutrients):
  - ❒ Using species of varying heights to optimize light distribution.
  - ❒ Staggering maturity times to reduce light competition.
  - ❒ Employing pruning to regulate canopy density.
- ☆ **Temporal management** ensures maximum resource use throughout the year:
  - ❒ Selecting species with diverse phenologies and life cycles.
  - ❒ Implementing relay planting and staggered sowing.

Initial Tree Spacing

Spacing should consider:

- ☆ Tree and crop growth rates.
- ☆ Tree-crop value ratios.
- ☆ Environmental resource availability (light, water, nutrients).
- ☆ Machinery access and ease of management.
- ☆ Latitude and topography (which influence solar radiation).
- ☆ Farmer preferences and system objectives.

Key Agroforestry Management Practices

1. Thinning

This practice involves selective removal of immature or inferior trees to:

- ☆ Improve health and vigor of remaining trees.
- ☆ Enhance light penetration and space for growth.
- ☆ Reduce disease and pest incidence.
- ☆ Increase net yield and timber quality.

2. Pruning

Pruning involves the removal of branches to:

- ☆ Improve timber quality (clear, straight grain).
- ☆ Facilitate crop growth by allowing more light.
- ☆ Control tree shape and canopy size.
- ☆ It can be natural (self-pruning) or artificial (manual cutting).

3. Pollarding

Pollarding is the cutting of trees at a height of 2–2.5 meters to promote new shoots, especially used in species like *Salix, Hardwickia, Grewia,* and *Morus.*

4. Lopping

Lopping refers to the removal of fresh shoots or leaves from the tree crown for fodder. It is common in *Morus, Grewia,* and *Bauhinia* species.

5. Coppicing

Coppicing involves cutting the main stem at 20–30 cm above ground to allow regrowth from the stump. Species vary in their coppicing ability:

- ✰ Strong coppicers: *Acacia catechu, Albizia lebbeck.*
- ✰ Good coppicers: *Aesculus indica, Chloroxylon swietenia.*
- ✰ Poor coppicers: *Adina cordifolia, Bombax ceiba.*
- ✰ Non-coppicers: Most coniferous trees.

6. Bending

Bending is practiced to suppress apical dominance and redirect energy towards lateral growth, thereby encouraging foliage production near ground level.

7. Training

Vertical training of tree stems in agroforestry ensures reduced shading and minimized competition for light with understory crops. This is particularly important in systems where light interception critically affects yield.

This comprehensive framework enables optimal selection and sustainable management of MPTS in agroforestry systems, ensuring improved productivity, ecological balance, and farmer livelihood.

QUESTION BANK

1. **Regeneration of tree species from seeds is known as:**
   A. Vegetative propagation B. Natural regeneration
   C. Artificial regeneration D. Coppicing

Answer: B. Natural regeneration

2. **Coppicing is regeneration from:**
   A. Roots B. Seeds
   C. Stump or base of tree after felling D. Rhizomes

Answer: C. Stump or base of the tree after felling

3. **Which of the following species is known for good coppicing ability?**
   A. Dalbergia sissoo B. Populus deltoides
   C. Tectona grandis D. Acacia nilotica

Answer: A. Dalbergia sissoo

4. **Root suckers arise from:**
   A. Branches B. Leaves
   C. Adventitious buds on roots D. Terminal buds

Answer: C. Adventitious buds on roots

5. **Transplanting of seedlings is typically done to:**
   A. Increase seed dormancy
   B. Change genetic traits
   C. Facilitate root development and survival
   D. Induce flowering

Answer: C. Facilitate root development and survival

6. **Stump planting is commonly used in propagation of:**
   A. Poplar B. Teak
   C. Leucaena leucocephala D. Mango

Answer: C. Leucaena leucocephala

7. Branch cutting as a vegetative propagation method is not suitable for:

A. Populus spp.
B. Eucalyptus spp.
C. Mango
D. Gliricidia sepium

Answer: C. Mango

8. Which structure is used for regeneration in rhizome-based propagation?

A. Horizontal underground stem
B. Aerial roots
C. Tap root
D. Leaf petiole

Answer: A. Horizontal underground stem

9. An advantage of propagation by rhizomes is:

A. Slow growth
B. Uniformity in progeny
C. Genetic variation
D. Low survival rate

Answer: B. Uniformity in progeny

10. Which of the following trees regenerates well from root suckers?

A. Shisham (*Dalbergia sissoo*)
B. Sal (*Shorea robusta*)
C. Neem (*Azadirachta indica*)
D. Teak (*Tectona grandis*)

Answer: B. Sal (*Shorea robusta*)

Chapter 14

Nursery Bed Preparation and Management

Nursery bed preparation and management is a fundamental step in raising healthy and vigorous seedlings for forestry and agroforestry plantations. The process begins with the selection of a suitable site that has good soil drainage, ample sunlight, and access to water. The land is then cleared of weeds, stones, and debris, followed by thorough ploughing and leveling to create a fine tilth for sowing. Raised nursery beds, typically 1 meter wide and of convenient length, are prepared to improve drainage and avoid waterlogging. These beds are usually oriented east to west to ensure uniform sunlight distribution.

The soil used in nursery beds should be a well-drained sandy loam, rich in organic matter, and free from pathogens. To enhance fertility and structure, compost or well-rotted farmyard manure is mixed into the topsoil. In some cases, soil sterilization techniques like solarization, steaming, or application of fungicides are used to prevent soil-borne diseases. Seeds are then sown directly or in lines with appropriate spacing to facilitate thinning and weeding later. Depending on the species, seeds may require pre-treatment such as soaking or scarification to enhance germination.

After sowing, seeds are lightly covered with fine soil or sand and the beds are irrigated gently using watering cans or sprinkler systems. Regular watering is essential to keep the soil moist, especially during germination and early seedling stages. Shade is provided using thatch, cloth, or shade nets to protect delicate seedlings from direct sunlight, which is gradually removed during the hardening phase. Regular maintenance includes weeding, pest and disease control, thinning of overcrowded seedlings, and pricking out where necessary to avoid competition and promote uniform growth.

Hardening-off is a critical stage in which seedlings are gradually exposed to natural conditions to prepare them for transplanting in the main field. Throughout the nursery period, proper labeling and record-keeping of each bed help in monitoring seedling development and planning further activities. Effective nursery bed preparation and management not only improve the survival and performance of seedlings but also ensure the long-term success of planting programs in forestry and agroforestry systems.

Definition of Nursery and Related Terms

A **nursery** is defined as a designated area where plants, especially tree seedlings, are raised for eventual transplantation in the field. It serves as the foundation for successful plantation and afforestation programs.

- A **nursery bed** is a specially prepared area within the nursery used for sowing seeds or placing transplants and cuttings. Nursery beds are typically classified into two categories based on their function:
  - **Seedling Beds**: Where seeds are sown and seedlings are initially raised.
  - **Transplant Beds**: Where seedlings from seedling beds are transplanted and hardened before field planting.
- A **Seedling Nursery** includes only seedling beds and does not involve transplanting operations.
- A **Transplant Nursery** contains only transplant beds, where seedlings are prepared for final plantation.
- In most Indian nurseries, both seedling and transplant beds are often maintained in the same area.
- The term **nursery stock** refers to all plants raised and maintained in the nursery for future planting.

Importance and Objectives of Nursery Establishment

Nurseries play a vital role in **artificial regeneration** and the success of **plantation forestry**, especially in the propagation of fast-growing or exotic species. The importance of nursery practices can be summarized as follows:

1. **Irregular Seed Production**: Many tree species do not produce viable seeds annually. Nurseries allow for the propagation of such species by sowing seeds collected during good seed years and maintaining seedlings for use over subsequent years.
2. **Protection of Slow-Growing Species**: Direct sowing in the field often results in poor survival due to weed competition. Nurseries provide a controlled environment for such species to grow until they are robust enough to be transplanted.

3. **Avenue Plantations**: For roadside and ornamental plantations, tall and sturdy seedlings are essential, which can only be raised in nursery conditions.
4. **Higher Establishment Success**: Some species perform poorly when raised by direct sowing but show excellent results when planted as nursery-grown seedlings.
5. **Introduction of Exotic Species**: Species such as tropical pines, poplars, and eucalyptus are best introduced through nursery propagation.
6. **Afforestation of Marginal Lands**: Degraded or barren sites are best regenerated using robust nursery-raised planting material.
7. **Replacement of Casualties**: Mortality in plantations is inevitable. Effective replanting during the same or subsequent season is only possible using nursery-raised stock, as direct sowing fails due to weed pressure and delayed establishment.

Types of Nurseries

Nurseries may be classified based on **irrigation availability** or **duration of use**:

A. Based on Irrigation Availability

1. **Dry Nursery**: Maintained without any irrigation or artificial watering. Seedling survival depends entirely on natural precipitation.
2. **Wet Nursery**: Maintained with regular irrigation or artificial watering during dry periods to ensure optimal seedling growth.

B. Based on Duration of Use

1. Temporary Nursery

- ☆ Established for short-term use, typically in or near plantation sites.
- ☆ Also referred to as a **field nursery**, it caters to small-scale or localized planting needs.

Advantages

- ☆ Often established on freshly cleared land rich in humus, minimizing the need for additional manuring.
- ☆ Reduced pest, disease, and weed problems.
- ☆ Ideal for raising species in their optimal altitudinal zones in hilly areas.
- ☆ Proximity to planting site reduces transportation cost and transplanting shock.

Disadvantages

- ✰ Lack of proper supervision due to remote location.
- ✰ Absence of irrigation facilities can result in poor seedling growth and high mortality.
- ✰ Comparatively expensive to maintain and less efficient than permanent nurseries.

2. Permanent Nursery

- ✰ Maintained continuously to supply planting stock over a long duration.
- ✰ Designed to serve multiple plantation programs and large areas; usually located near administrative headquarters or accessible roads for ease of monitoring and transport.

Features

- ✰ Equipped with irrigation, shade provision, and storage facilities.
- ✰ Requires regular fertilization as the soil fertility declines over time due to continuous usage.
- ✰ Better management and consistent seedling quality.

Criteria for Nursery Site Selection

Selecting an appropriate site is crucial for efficient nursery operation and seedling production. The following factors should be considered:

1. **Proximity to Planting Area**: The site should be chosen with reference to the location of proposed plantations to minimize transport costs and transplanting shock.
2. **Accessibility and Supervision**: The nursery should be easily accessible to ensure regular supervision and timely operations.
3. **Availability of Water**: The site should be close to a reliable water source, preferably located slightly below it to allow gravity-fed irrigation.
4. **Soil Quality**: Ideal soil is well-drained **sandy loam** with good aeration. Clayey soils should be avoided due to poor drainage and their tendency to crack during dry seasons.
5. **Topography**: In hilly regions, gentle slopes (less than 5°) are preferred. Steeper slopes require terracing, which increases costs.
6. **Aspect**:
   - ❒ In the Western Himalayas, **northern aspects** are generally preferable due to milder conditions.

- For higher elevations (above 1200 m), **western or southwestern aspects** are more suitable as southern slopes become excessively cold and wet.

7. **Altitudinal Suitability**: The selected site should fall within the optimal **altitudinal range** of the species intended to be raised.
8. **Frost-Free Zones**: Areas prone to frost should be avoided as young seedlings are highly susceptible to frost damage.

QUESTION BANK

1. **What is the ideal width of a standard nursery bed?**

   A. 2 meters
   B. 0.5 meters
   C. 1 meter
   D. 3 meters

Answer: C. 1 meter

2. **What is the main purpose of raising nursery beds above ground level?**

   A. To increase shade
   B. To reduce water loss
   C. To improve drainage and avoid waterlogging
   D. To increase temperature

Answer: C. To improve drainage and avoid waterlogging

3. **Which type of soil is most suitable for nursery bed preparation?**

   A. Sandy soil
   B. Clayey soil
   C. Rocky soil
   D. Sandy loam soil

Answer: D. Sandy loam soil

4. **What is the ideal orientation of nursery beds for uniform sunlight?**

   A. North–South
   B. East–West
   C. Northeast–Southwest
   D. Northwest–Southeast

Answer: B. East–West

5. **Why is organic manure mixed in the nursery bed soil?**

   A. To reduce soil weight
   B. To add color to the bed
   C. To increase fertility and improve structure
   D. To attract insects

Answer: C. To increase fertility and improve structure

6. **Which of the following is used to sterilize the nursery bed soil?**

   A. Urea
   B. Cow dung
   C. Steaming or solarization
   D. Charcoal

Answer: C. Steaming or solarization

7. **Which tool is commonly used for gentle irrigation in nursery beds?**
   A. Hose pipe B. Sprinkler
   C. Watering can D. Bucket

Answer: C. Watering can

8. **What is the function of providing shade over nursery beds?**
   A. To avoid pest attack
   B. To reduce seedling competition
   C. To protect seedlings from direct sunlight
   D. To slow down germination

Answer: C. To protect seedlings from direct sunlight

9. **What is the term for gradually exposing nursery seedlings to outdoor conditions?**
   A. Transplanting B. Mulching
   C. Pricking D. Hardening-off

Answer: D. Hardening-off

10. **Which of the following is a recommended maintenance activity in nursery beds?**
    A. Ploughing B. Water harvesting
    C. Weeding and thinning D. Broadcasting fertilizer

Answer: C. Weeding and thinning

Chapter 15

Cultural Practices for Bare Root and Seedling, Field Handling of Nursery Stock

Nurturing the Future Forest: Cultural Practices for Bare Root and Seedling Handling in the Field

The success of any reforestation or afforestation effort hinges on the careful handling and planting of nursery stock. Bare root and seedling plants, representing the future of our forests, require meticulous attention from the nursery to the field to ensure their survival and thriving. This article delves into the critical cultural practices for handling these young trees, focusing on minimizing stress and maximizing establishment rates.

Understanding Bare Root and Seedling Vulnerability

Bare root and seedling plants, while cost-effective and diverse, are particularly vulnerable during transport and planting. Unlike containerized plants, they lack the protective root ball of soil. This makes them susceptible to:

- ☆ **Desiccation:** Roots dry out quickly when exposed to air, leading to root damage and reduced survival.
- ☆ **Temperature Extremes:** Both freezing and excessive heat can damage tender roots and shoots.
- ☆ **Physical Damage:** Rough handling can break roots and stems, hindering growth and increasing vulnerability to disease.
- ☆ **Planting Shock:** The sudden change in environment from the nursery to the field can be stressful.

Therefore, implementing proper cultural practices is paramount to minimizing these risks and ensuring successful establishment.

I. Nursery Stock Preparation and Handling

The journey to a healthy forest begins at the nursery. Implementing the following practices ensures the best possible start for the young trees:

- ✰ **Grading and Culling:** Thoroughly inspect all seedlings upon arrival. Reject any with signs of disease, damage, or poor root development. Consistently grading and culling ensures resources are focused on the healthiest and most likely to succeed plants.
- ✰ **Proper Storage:** Store bare root seedlings in a cool, dark, and humid environment. Ideal storage conditions include temperatures near freezing (33-38°F or 0.5-3°C) and high humidity (above 90 per cent). Refrigerated storage is ideal. If refrigeration is unavailable, consider heeling-in the seedlings in a shaded area, covering the roots with moist soil or mulch.
- ✰ **Minimizing Storage Time:** Aim to plant seedlings as soon as possible after receiving them. Prolonged storage increases the risk of desiccation and disease.
- ✰ **Hydration:** Before transport, ensure seedlings are adequately hydrated. Gentle soaking of roots for a short period can help, but avoid prolonged soaking which can lead to anaerobic conditions and root rot.

II. Field Handling and Planting Techniques

The transition from nursery to field is a critical phase. Minimizing stress during this period is crucial for successful plant establishment.

- ✰ **Transportation:** Transport seedlings to the planting site in a manner that protects them from desiccation, temperature extremes, and physical damage. Cover seedling bundles with tarps or blankets during transport. Use refrigerated trucks if transporting over long distances.
- ✰ **On-Site Storage:** At the planting site, continue to protect seedlings from the elements. Store them in a shaded area and keep the roots moist. Avoid stacking bundles too high, as this can restrict airflow and lead to overheating.
- ✰ **Planting Time:** Choose planting times that minimize stress. For bare root seedlings, early spring or late fall are generally preferred when temperatures are cooler and moisture is more abundant. Avoid planting during periods of extreme heat, drought, or heavy frost.
- ✰ **Planting Depth:** Plant seedlings at the correct depth. The root collar (the point where the stem transitions into the roots) should be level

with the soil surface. Planting too deep can suffocate the roots, while planting too shallow can expose them to drying.

- ☆ **Root Pruning:** Prior to planting, prune any damaged or excessively long roots. This encourages new root growth and improves the seedling's ability to establish in the new environment.
- ☆ **Planting Hole:** Dig a hole that is large enough to accommodate the seedling's root system without crowding or bending.
- ☆ **Proper Backfilling:** Carefully backfill the planting hole with soil, ensuring that the roots are in contact with the soil. Gently tamp the soil around the base of the seedling to eliminate air pockets.
- ☆ **Watering:** Immediately after planting, water seedlings thoroughly to settle the soil and provide essential moisture.
- ☆ **Mulching:** Applying a layer of mulch around the base of the seedling can help conserve moisture, suppress weed growth, and moderate soil temperature. Organic mulches like wood chips or straw are ideal.
- ☆ **Protection:** Consider using tree shelters or browsing guards to protect seedlings from animal damage.
- ☆ **Handling with Care:** Emphasize gentle handling throughout the entire planting process. Avoid dropping or throwing seedlings, as this can damage the roots and reduce their chances of survival.

III. Cultural Practices to Support Establishment

Even with careful handling and planting, ongoing cultural practices are often necessary to support seedling establishment and growth:

- ☆ **Weed Control:** Competing vegetation can rob seedlings of water, nutrients, and sunlight. Implement weed control measures, such as manual weeding, herbicide application, or mulching, to create a favorable growing environment.
- ☆ **Watering:** In areas with limited rainfall, supplemental watering may be necessary during the first few growing seasons, especially during periods of drought.
- ☆ **Fertilization:** If the soil is deficient in essential nutrients, consider applying a balanced fertilizer to promote growth. Soil testing can help determine the specific nutrient needs of the seedlings.
- ☆ **Pest and Disease Management:** Regularly monitor seedlings for signs of pests and diseases. Implement appropriate control measures to prevent or mitigate damage.
- ☆ **Pruning and Thinning:** As the trees grow, pruning and thinning may be necessary to improve stand health and promote timber quality.

The successful establishment of bare root and seedling plants requires careful planning, diligent execution, and a commitment to proper cultural practices. By understanding the vulnerabilities of these young trees and implementing the techniques outlined in this article, we can significantly increase their survival rates and ensure the healthy growth of future forests. Investing in these seemingly small details pays massive dividends in the long run, guaranteeing a more sustainable and resilient future for our planet.

QUESTION BANK

1. **What is the key requirement before lifting bare-root seedlings from nursery beds?**
   A. Complete drying of the soil
   B. Heavy irrigation
   C. Light irrigation or soil moist condition
   D. Root pruning

Answer: C. Light irrigation or soil moist condition

2. **What is the main advantage of using bare-root seedlings?**
   A. Heavier transport
   B. Higher water requirement
   C. Cheaper and easier to transport
   D. Longer establishment time

Answer: C. Cheaper and easier to transport

3. **Which of the following is not a recommended practice for handling seedlings during transplanting?**
   A. Exposing roots to direct sunlight
   B. Keeping roots moist
   C. Planting during cooler hours
   D. Trimming oversized roots

Answer: A. Exposing roots to direct sunlight

4. **In field planting, what is the ideal time to transplant nursery seedlings?**
   A. During peak noon
   B. Midnight
   C. Early morning or late afternoon
   D. Mid-winter

Answer: C. Early morning or late afternoon

5. **Which of the following tools is commonly used for lifting bare-root seedlings?**
   A. Plough
   B. Hoe
   C. Digging fork or spade
   D. Rake

Answer: C. Digging fork or spade

6. **What is the role of root pruning in seedling management?**
   A. To increase shoot height only
   B. To promote fibrous root development
   C. To reduce nutrient uptake
   D. To delay transplanting

Answer: B. To promote fibrous root development

7. What should be done immediately after uprooting bare-root seedlings for field transport?

A. Dip in herbicide

B. Expose to air to harden

C. Pack with moist material or wrap roots

D. Wash roots clean

Answer: C. Pack with moist material or wrap roots

8. The best stage for transplanting seedlings to the field is when they have:

A. Two or fewer leaves

B. Very long roots

C. Well-developed root system and strong shoot

D. Yellowing leaves

Answer: C. Well-developed root system and strong shoot

9. Which of the following is most important to avoid transplant shock in seedlings?

A. Heavy watering after one week

B. Keeping seedlings in the sun before transplanting

C. Proper root moisture and minimal root damage

D. Applying urea before planting

Answer: C. Proper root moisture and minimal root damage

10. What is "heeling-in" in nursery management?

A. Removing diseased roots

B. Storing seedlings temporarily in moist trenches

C. Dipping roots in fertilizer

D. Root cutting before transport

Answer: B. Storing seedlings temporarily in moist trenches

Chapter 16

Silviculture of Important Tree Species, Choice of Species: Site Factors, Root, Crown and Bole Characteristics, Phenology, Nutritional and Water Requirement, Ground Operation, Tending, Harvesting Utility

The **silviculture of important tree species** involves understanding their biological and ecological characteristics to ensure successful growth, productivity, and sustainability in different agroforestry systems. **Choosing the right species** is crucial and depends on several **site factors** such as soil type, topography, rainfall, temperature, and altitude. For instance, drought-tolerant species like *Prosopis juliflora* and *Acacia nilotica* are suited to arid zones, whereas species like *Tectona grandis* and *Gmelina arborea* thrive in moist, well-drained tropical areas. The **root system** of a tree species—whether deep taproot or shallow lateral roots—determines its adaptability to moisture regimes and compatibility with crops. Trees with deep taproots are generally preferred in agroforestry as they do not compete heavily with crops for nutrients and water. Similarly, **crown architecture**—open, sparse or dense—impacts light availability to the understorey crops; species with a light-filtering or sparse crown like *Albizia lebbeck* or *Leucaena leucocephala* are preferred for intercropping. **Bole characteristics** such as straightness, height, and timber density influence harvesting ease and economic value. The **phenology** of tree species—timing of leaf flush, flowering, and fruiting—helps determine planting and pruning schedules and avoid competition during critical crop stages. Understanding the

nutritional requirements is vital; leguminous trees like *Gliricidia sepium* and *Sesbania* enrich soil fertility through nitrogen fixation, while fast-growing species often demand higher nutrient inputs. Similarly, **water requirement** varies—some trees are suited to dryland systems while others need frequent watering or high rainfall. **Ground operations** like land preparation, pitting, manuring, and planting must align with the biological needs of the species. **Tending operations** such as weeding, thinning, pruning, coppicing, and protection from pests and fire ensure healthy stand development. At maturity, appropriate **harvesting techniques** like selective felling or clear-cutting are employed based on the species and intended use. Finally, the **utility** of tree species spans multiple domains—providing timber, fuelwood, fodder, fiber, medicinal products, and ecological services like erosion control, carbon sequestration, and biodiversity conservation. Thus, a sound understanding of silvicultural practices enhances productivity and ecological harmony in agroforestry systems.

Silviculture of Important Tree Species: A Comprehensive Guide

Silviculture, the practice of controlling the establishment, growth, composition, health, and quality of forests to meet diverse needs and values, is a critical aspect of forest management. The choice of tree species for a particular site is a crucial decision that a silviculturist must make. This article provides an in-depth analysis of the silviculture of important tree species, including site factors, root, crown, and bole characteristics, phenology, nutritional and water requirements, ground operations, tending, and harvesting utility.

Site Factors

The selection of tree species depends on various site factors such as soil type, topography, climate, and vegetation. The site factors determine the growth and yield of trees, and the suitability of a particular species for a site. For instance, some species are well-adapted to sandy soils, while others prefer clay or loamy soils. The slope and aspect of the site also influence the choice of species, as some species can tolerate steep slopes and shade, while others require full sun and level ground.

Root, Crown, and Bole Characteristics

The root, crown, and bole characteristics of a tree species are essential factors to consider in silviculture. The root system of a tree affects its stability, water and nutrient uptake, and competition with other plants. Species with deep and wide-spreading root systems are suitable for sites with shallow soils or high wind exposure. The crown characteristics, such as the shape, size, and density, influence the tree's light requirements, shade tolerance, and susceptibility to windthrow. The bole characteristics, such as the diameter, height, and form, determine the tree's timber value, strength, and durability.

Phenology

Phenology, the study of periodic plant and animal life cycle events and how these are influenced by seasonal and interannual variations in climate, is an essential factor in silviculture. The phenology of a tree species, such as its budding, flowering, and leaf-shedding patterns, affects its growth and development, susceptibility to pests and diseases, and compatibility with other species. For instance, early-successional species, which grow rapidly and have short lifespans, have different phenological characteristics than late-successional species, which grow slowly and have long lifespans.

Nutritional and Water Requirements

The nutritional and water requirements of a tree species are critical factors in silviculture. Species that require high nutrient and water inputs may not be suitable for nutrient-poor or dry sites. Conversely, species that are drought-tolerant and have low nutrient requirements may be suitable for sites with low rainfall or nutrient availability. The silvicultural practices, such as thinning, pruning, and fertilization, can influence the nutrient and water availability for trees.

Ground Operations, Tending, and Harvesting Utility

Ground operations, such as site preparation, planting, and tending, are essential silvicultural practices that influence the establishment, growth, and survival of tree species. The choice of species depends on the ground operation requirements, such as the planting density, spacing, and machine accessibility. Tending practices, such as thinning, pruning, and cleaning, are necessary to promote the growth and quality of trees and to reduce the competition among them. The harvesting utility of a tree species, such as its timber value, wood properties, and market demand, is an essential factor in silviculture. The harvesting practices, such as the cutting method, season, and frequency, depend on the tree species and the intended use of the wood.

The silviculture of important tree species requires a comprehensive understanding of the site factors, root, crown, and bole characteristics, phenology, nutritional and water requirements, ground operations, tending, and harvesting utility. The choice of species for a particular site depends on various factors, and the silvicultural practices must be tailored to the species and site conditions. A successful silvicultural program requires careful planning, monitoring, and adaptation to changing conditions and objectives.

QUESTION BANK

1. **Which of the following species is best suited for planting in arid and semi-arid regions due to its deep taproot system and drought tolerance?**

   A. *Tectona grandis* B. *Dalbergia sissoo*

   C. *Acacia nilotica* D. *Populus deltoides*

Answer: C. *Acacia nilotica*

2. **The selection of tree species for a site is primarily based on which of the following factors?**

   A. Timber price only

   B. Farmer's choice only

   C. Site characteristics like soil, climate, and rainfall

   D. Availability of government subsidy

Answer: C. Site characteristics like soil, climate, and rainfall

3. **Tree species with a sparse crown are preferred in agroforestry because:**

   A. They require less water

   B. They grow faster

   C. They allow sunlight to reach the crops below

   D. They are easier to prune

Answer: C. They allow sunlight to reach the crops below

4. **Which tree species is known for its nitrogen-fixing ability and is often used to improve soil fertility in agroforestry systems?**

   A. *Eucalyptus globulus* B. *Gmelina arborea*

   C. *Leucaena leucocephala* D. *Pinus roxburghii*

Answer: C. *Leucaena leucocephala*

5. **Phenology in silviculture refers to:**

   A. The pruning schedule of trees

   B. Study of root system architecture

   C. The seasonal patterns of leafing, flowering, and fruiting

   D. Tree harvesting intervals

Answer: C. The seasonal patterns of leafing, flowering, and fruiting

6. **Which of the following species is commonly used in agroforestry for its fast growth, coppicing ability, and good fuelwood quality?**
   A. *Shorea robusta*
   B. *Terminalia arjuna*
   C. *Eucalyptus tereticornis*
   D. *Mangifera indica*

Answer: C. *Eucalyptus tereticornis*

7. **A tree with a straight and cylindrical bole is considered ideal for:**
   A. Firewood
   B. Fodder production
   C. High-quality timber
   D. Soil erosion control

Answer: C. High-quality timber

8. **Which of the following is a ground operation carried out before tree planting?**
   A. Thinning
   B. Coppicing
   C. Site preparation and pitting
   D. Pruning

Answer: C. Site preparation and pitting

9. **Tending operations in silviculture include all the following EXCEPT:**
   A. Weeding
   B. Pruning
   C. Harvesting
   D. Thinning

Answer: C. Harvesting

10. **In water-scarce regions, tree species selected for planting should ideally:**
   A. Have large, shallow roots
   B. Be evergreen and high water demanding
   C. Have deep root systems and low water requirements
   D. Be deciduous with shallow roots

Answer: C. Have deep root systems and low water requirements

Chapter 17

Horticulture and Forage Crops-Based Agroforestry Models Developed by ICAR-IGFRI

Introduction

The Indian Council of Agricultural Research-Indian Grassland and Fodder Research Institute (ICAR-IGFRI) has been at the forefront of developing innovative agroforestry models that incorporate horticulture and forage crops. These models have the potential to revolutionize the way we think about agriculture and land use, offering a more sustainable and profitable approach for farmers.

Agroforestry: A Sustainable Approach

Agroforestry is a land use management system that combines trees and/or shrubs with crops and/or livestock on the same piece of land. This approach has numerous benefits, including increased biodiversity, improved soil health, reduced erosion, and enhanced carbon sequestration. Additionally, agroforestry can provide farmers with multiple sources of income, as well as products such as fruit, nuts, and fodder.

Horticulture and Forage Crops in Agroforestry

ICAR-IGFRI has been developing agroforestry models that incorporate horticulture and forage crops. Horticulture crops, such as fruits, vegetables, and nuts, can provide farmers with a valuable source of income, as well as nutrition for their families. Forage crops, on the other hand, can be used to feed livestock, providing another source of income for farmers.

The institute has developed 17 different agroforestry models that incorporate horticulture and forage crops. These models have been tested in various agro-climatic zones across India, and have been found to be highly effective in increasing crop yields, improving soil health, and providing farmers with multiple sources of income.

One such model is the "Horti-Fodder" model, which combines fruit trees with forage crops. This model has been found to be particularly effective in dryland areas, where water is scarce. The fruit trees provide shade for the forage crops, which helps to reduce water evaporation and conserve moisture in the soil. Additionally, the forage crops provide a source of nutrition for livestock, which can help to increase milk yields and provide another source of income for farmers.

Another model developed by ICAR-IGFRI is the "Agri-Horti-Silvi" model, which combines crops, fruits, and timber trees. This model has been found to be highly effective in increasing crop yields, improving soil health, and providing farmers with multiple sources of income. The timber trees can be harvested for wood, providing yet another source of income for farmers.

Commonly Adopted Horticulture and Forage Crops-Based Agroforestry Models

Horticulture and Forage-Based Agroforestry Models

1. **Agri-Horti System**
   - **Components:** Agricultural crops + Horticultural trees
   - **Example crops:**
     - **Horticultural:** Mango (*Mangifera indica*), Guava (*Psidium guajava*), Amla (*Emblica officinalis*), Ber (*Zizyphus mauritiana*)
     - **Agricultural:** Pulses (*e.g.*, pigeon pea, black gram), vegetables (*e.g.*, okra, brinjal)
   - **Suitable regions:** Semi-arid and arid regions
   - **Benefits:** Diversification of income, improved soil fertility, better land use efficiency

2. **Horti-Pastoral System**
   - **Components:** Horticultural trees + Pasture/Forage grasses
   - **Example crops:**
     - **Horticultural:** Pomegranate (*Punica granatum*), Citrus (*Citrus spp.*), Custard apple (*Annona squamosa*)
     - **Forage:** Napier grass (*Pennisetum purpureum*), Guinea grass (*Panicum maximum*), Stylosanthes spp.

- **Suitable regions:** Rainfed zones, degraded lands
- **Benefits:** Enhanced fodder availability, soil conservation, orchard-based livestock support

3. Silvi-Horti System

- **Components:** Forest tree species + Fruit trees
- **Examples:**
  - **Silvicultural trees:** Teak (*Tectona grandis*), Subabul (*Leucaena leucocephala*), Neem (*Azadirachta indica*)
  - **Horticultural crops:** Sapota (*Manilkara zapota*), Lime (*Citrus aurantifolia*)
- **Use:** Promotes both fruit production and timber/firewood yield
- **Best for:** Marginal lands and boundary plantations

4. Horti-Agri-Pastoral System (Multicomponent)

- **Components:** Fruit trees + Field crops + Forage/pasture
- **Example configuration:**
  - **Fruits:** Mango, Papaya
  - **Field crops:** Groundnut, Soybean, Maize
  - **Fodder:** Cowpea, Lucerne, or Stylosanthes
- **Benefits:** Sustainable use of land, diversified output, soil fertility improvement

5. Fodder-Fruit Tree Model

- **Components:** Fodder trees interplanted with fruit trees
- **Example species:**
  - **Fodder trees:** Subabul, Gliricidia, Sesbania
  - **Fruits:** Banana, Guava, Custard apple
- **Application:** Used where livestock and orchard-based income are needed simultaneously

6. Forage Alley Cropping with Horticultural Boundary Plantation

- **Components:** Fodder crops grown in alleys between rows of fruit trees
- **Example:**
  - **Fodder alleys:** Hybrid Napier, Lucerne
  - **Boundary horticulture:** Papaya, Citrus, Pomegranate
- **Advantages:** Maximized light use, erosion control, dual income from fruit and fodder

7. Homestead Agroforestry (Home Garden System)

- ✰ **Components:** Mixed horticultural crops + Forage grasses + Timber/fuelwood trees + Vegetables
- ✰ **Species mix:** Coconut, Banana, Papaya, Taro, Jackfruit, Guava, plus forage grasses like Cowpea
- ✰ **Common in:** Kerala, Northeastern states, Coastal India
- ✰ **Function:** Subsistence farming system with high diversity and ecological balance

Benefits of Horticulture and Forage Crops-based Agroforestry Models

The horticulture and forage crops-based agroforestry models developed by ICAR-IGFRI offer numerous benefits for farmers, including:

1. Increased Crop Yields: The models have been found to increase crop yields, providing farmers with more produce to sell or consume.
2. Improved Soil Health: The models help to improve soil health, which can lead to higher crop yields in the long term.
3. Multiple Sources of Income: The models provide farmers with multiple sources of income, including from crops, fruits, nuts, and livestock.
4. Enhanced Biodiversity: The models help to increase biodiversity, which can lead to a more resilient agricultural system.
5. Reduced Erosion: The models help to reduce erosion, which can help to preserve soil health and prevent nutrient loss.
6. Carbon Sequestration: The models can help to sequester carbon, which can help to mitigate climate change.

The horticulture and forage crops-based agroforestry models developed by ICAR-IGFRI offer a promising approach to sustainable agriculture. By combining crops, fruits, nuts, and livestock, these models provide farmers with multiple sources of income, improve soil health, and increase biodiversity. As the world grapples with the challenges of climate change, food security, and rural poverty, these models offer a valuable tool for building more resilient and sustainable agricultural systems.

QUESTION BANK

1. **Which institute has developed several agroforestry models involving horticulture and forage crops to enhance forage-livestock productivity?**

   A. ICRISAT B. ICFRE

   C. ICAR-IGFRI D. ICAR-NBPGR

Answer: C. ICAR-IGFRI

2. **In the Horti-pastoral model developed by IGFRI, which fruit tree is commonly intercropped with Guinea grass?**

   A. Mango B. Ber

   C. Pomegranate D. Amla

Answer: B. Ber

3. **The primary goal of IGFRI's forage-based agroforestry models is to:**

   A. Enhance timber production

   B. Promote exotic species

   C. Improve year-round green fodder availability

   D. Promote cash crops only

Answer: C. Improve year-round green fodder availability

4. **Which of the following is a forage crop commonly used in IGFRI's agroforestry models?**

   A. Bajra B. Napier grass

   C. Paddy D. Mustard

Answer: B. Napier grass

5. **The Horti-pastoral system promoted by IGFRI is best suited for which type of area?**

   A. Coastal sandy soils B. Arid and semi-arid regions

   C. Alpine highlands D. Urban landscapes

Answer: B. Arid and semi-arid regions

6. **Which of the following is NOT typically a component in IGFRI's Horti-pastoral agroforestry model?**

   A. Fruit tree B. Pulse crop

   C. Fodder grass D. Leguminous shrub

Answer: B. Pulse crop

7. **Which of the following combinations is correct in IGFRI's Agri-horti-pastoral model?**
   A. Mustard + Teak + Cowpea
   B. Maize + Amla + Stylosanthes
   C. Wheat + Neem + Sorghum
   D. Cotton + Bamboo + Berseem

Answer: B. Maize + Amla + Stylosanthes

8. **IGFRI's Horti-pastoral model involving Aonla and Stylosanthes aims to enhance:**
   A. Oil production
   B. Timber quality
   C. Soil nitrogen content and fodder supply
   D. Insect resistance

Answer: C. Soil nitrogen content and fodder supply

9. **Which ICAR institute is considered a pioneer in developing forage-based agroforestry models in India?**
   A. ICAR-CAZRI
   B. ICAR-CRIDA
   C. ICAR-IGFRI
   D. ICAR-IARI

Answer: C. ICAR-IGFRI

10. **The integration of horticulture with forage crops in agroforestry systems primarily contributes to:**
   A. Increased pesticide usage
   B. Monoculture promotion
   C. Livestock productivity and diversified income
   D. Exclusive timber production

Answer: C. Livestock productivity and diversified income

Chapter 18

Agroforestry Models Developed by Indian Council of Forestry Research and Education

Objective: Agroforestry Models Developed by the Indian Council of Forestry Research and Education (ICFRE) as a Path to Sustainable Land Management

The Indian Council of Forestry Research and Education (ICFRE) plays a pivotal role in advancing forestry research and education in India. Recognizing the crucial link between agriculture and forestry, ICFRE has developed a range of agroforestry models designed to promote sustainable land management, enhance livelihoods, and combat environmental degradation. These models offer a practical pathway for farmers and landowners to integrate trees and shrubs into their agricultural systems, creating mutually beneficial relationships.

Why Agroforestry?

Agroforestry offers a compelling alternative to conventional agriculture, particularly in regions facing climate change, soil degradation, and water scarcity. Integrating trees into agricultural landscapes provides numerous benefits, including:

- ☆ **Improved Soil Health:** Trees contribute organic matter to the soil, enhancing its structure, fertility, and water-holding capacity.
- ☆ **Increased Biodiversity:** Agroforestry systems provide habitat for a diverse range of flora and fauna, contributing to ecological balance.
- ☆ **Climate Change Mitigation:** Trees sequester carbon dioxide from the atmosphere, helping to reduce greenhouse gas emissions.

- **Diversified Income:** Farmers can generate income from a variety of products, including timber, fruits, nuts, fodder, and medicinal plants.
- **Enhanced Resilience:** Agroforestry systems are more resistant to extreme weather events like droughts and floods.

ICFRE's Role in Agroforestry Development

ICFRE's contributions to agroforestry in India are significant. Through its research institutions and extension programs, ICFRE:

- **Develops and Refines Agroforestry Models:** ICFRE research centers across the country develop region-specific agroforestry models tailored to local climate, soil conditions, and market demands.
- **Conducts Research on Species Selection:** Identifying suitable tree species for integration with different agricultural crops is crucial. ICFRE conducts research to determine the best tree-crop combinations.
- **Provides Training and Extension Services:** ICFRE provides training to farmers, extension workers, and other stakeholders on agroforestry techniques, promoting the adoption of these practices.
- **Develops Planting Material:** Ensuring the availability of high-quality seedlings of appropriate tree species is essential for successful agroforestry initiatives.
- **Promotes Policy Support:** ICFRE advocates for policies that support agroforestry development, including incentives for farmers and streamlined regulatory processes.

Examples of ICFRE-Developed Agroforestry Models

ICFRE has developed a multitude of agroforestry models, each designed to address specific regional challenges and opportunities. Some notable examples include:

- **Silvi-pastoral Systems:** Integrating trees with livestock farming, providing shade for animals, fodder resources, and improving pasture quality. Examples include *Acacia* species with grasses and leguminous forages.
- **Agri-silvicultural Systems:** Combining trees with annual or perennial crops, enhancing soil fertility and providing shade. Examples include *Populus deltoides* with wheat or sugarcane, and *Eucalyptus* with cotton.
- **Horti-silvicultural Systems:** Integrating fruit trees with timber trees or agricultural crops, providing multiple income streams. Examples include mango or guava with *Teak* or *Gamhar* and intercropped with vegetables.

- **Alley Cropping:** Planting trees in rows with agricultural crops grown in the alleys between the tree rows. This system is effective for soil conservation and improving crop yields.
- **Boundary Planting:** Planting trees along the boundaries of agricultural fields, providing windbreaks, reducing soil erosion, and generating timber resources.

Impact and Future Directions

ICFRE's agroforestry models have had a significant impact on rural livelihoods and environmental sustainability in India. By promoting the integration of trees into agricultural landscapes, these models have helped to enhance soil health, increase biodiversity, mitigate climate change, and improve the economic well-being of farmers.

Looking ahead, ICFRE continues to focus on:

- **Developing climate-resilient agroforestry models:** Adapting agroforestry practices to cope with the impacts of climate change.
- **Promoting participatory approaches:** Involving farmers in the design and implementation of agroforestry systems.
- **Strengthening market linkages:** Connecting farmers with markets for agroforestry products.
- **Utilizing technology:** Leveraging technology to improve the efficiency and effectiveness of agroforestry practices.
- **Expanding the scope of agroforestry:** Exploring new applications of agroforestry, such as urban agroforestry and agroforestry for carbon sequestration.

The agroforestry models developed by the Indian Council of Forestry Research and Education (ICFRE) offer a viable and sustainable pathway for land management in India. By promoting the integration of trees into agricultural systems, these models contribute to environmental conservation, economic development, and improved livelihoods for rural communities. As India faces the challenges of climate change and land degradation, ICFRE's continued efforts in agroforestry research and development will be crucial for building a more sustainable and resilient future.

QUESTION BANK

1. The Indian Council of Forestry Research and Education (ICFRE) is headquartered at:

A. Dehradun
B. New Delhi
C. Hyderabad
D. Jabalpur

Answer: A. Dehradun

2. ICFRE developed agroforestry models are mainly focused on:

A. Food grain intensification
B. Forestry research and tree improvement
C. Horticultural biotechnology
D. Integrated pest management

Answer: B. Forestry research and tree improvement

3. Which agroforestry model has been promoted by ICFRE for degraded land rehabilitation in central India?

A. Agrisilviculture
B. Silvipasture
C. Agrihorticulture
D. Bamboo-based system

Answer: A. Agrisilviculture

4. The ICFRE promotes the agri-silvi-horticulture model especially in:

A. Cold deserts
B. Arid and semi-arid regions
C. Coastal plains
D. Wet tropical zones

Answer: B. Arid and semi-arid regions

5. Which tree species is commonly used in ICFRE's agrisilviculture model in central India?

A. Pinus roxburghii
B. Eucalyptus tereticornis
C. Dalbergia sissoo
D. Abies pindrow

Answer: C. Dalbergia sissoo

6. The ICFRE model involving integration of grasses with trees for fodder and soil conservation is known as:

A. Agrihorticulture
B. Silvipasture
C. Agrisilviculture
D. Boundary plantation

Answer: B. Silvipasture

7. **One of the objectives of ICFRE agroforestry models is to:**
   A. Encourage monoculture plantations
   B. Minimize use of legumes
   C. Improve livelihood through sustainable forestry practices
   D. Promote urban vertical gardening

Answer: C. Improve livelihood through sustainable forestry practices

8. **ICFRE has promoted bamboo-based agroforestry models in which region of India?**
   A. Western Ghats
   B. North-Eastern India
   C. Indo-Gangetic Plains
   D. Rajasthan Plateau

Answer: B. North-Eastern India

9. **ICFRE's Agrihortisilviculture model is designed to integrate which three components?**
   A. Timber, oilseeds, and pulses
   B. Agriculture, horticulture, and forestry species
   C. Livestock, pulses, and cereals
   D. Soil, water, and biodiversity

Answer: B. Agriculture, horticulture, and forestry species

10. **Which of the following ICFRE institutes is actively involved in developing agroforestry models in arid regions?**
    A. Institute of Forest Biodiversity, Hyderabad
    B. Arid Forest Research Institute (AFRI), Jodhpur
    C. Tropical Forest Research Institute, Jabalpur
    D. Forest Research Institute (FRI), Dehradun

Answer: B. Arid Forest Research Institute (AFRI), Jodhpur

Introductory Agroforestry (Practical)

Experiment 1

Identification of Tree Species in Agroforestry

Objective: Identification of Seeds and Seedlings of MPT species

Identification of Tree Species in Agroforestry: Objective - Identification of Seeds and Seedlings of Multipurpose Tree Species

Preamble: Agroforestry is a sustainable land-use practice that integrates trees and shrubs with crops and/or livestock. Multipurpose tree species (MPTs) are vital components of agroforestry systems due to their diverse ecological and socio-economic benefits. Accurate identification of MPT seeds and seedlings is essential for the successful establishment and management of agroforestry systems. This illustration aims to provide an overview of the identification techniques for MPT seeds and seedlings in agroforestry systems.

Introduction: Agroforestry systems are designed to provide a range of ecological, economic, and social benefits. MPTs are key components of these systems due to their multiple uses, such as timber, fuelwood, fodder, and soil conservation. Accurate identification of MPT seeds and seedlings is crucial for the successful establishment and management of agroforestry systems. This illustration discusses various identification techniques for MPT seeds and seedlings, including morphological, molecular, and chemical methods.

Morphological Identification: Morphological identification involves the examination of physical characteristics, such as leaf shape, size, and color, as well as seed shape, size, and surface texture. Morphological identification can be challenging due to the high degree of variability within and among MPT species. To overcome this challenge, reference collections, field guides, and taxonomic keys can be used to aid in the identification process.

Molecular Identification: Molecular identification techniques involve the analysis of DNA sequences to determine the genetic identity of MPT seeds and seedlings. These methods are particularly useful when morphological identification is difficult or when distinguishing closely related species. Techniques such as DNA barcoding, polymerase chain reaction (PCR), and DNA sequencing can be employed for molecular identification of MPT seeds and seedlings.

Chemical Identification: Chemical identification involves the analysis of chemical compounds present in MPT seeds and seedlings. These compounds can be unique to specific species or groups of species and can serve as a basis for identification. Techniques such as gas chromatography (GC), high-performance liquid chromatography (HPLC), and mass spectrometry (MS) can be used for chemical identification of MPT seeds and seedlings.

Botanical Descriptions of Selected Multipurpose Tree Species

The botanical descriptions of selected multipurpose tree species, highlighting key characteristics of their fruits and seeds crucial for identification as below:

1. Mulberry (*Morus alba* L., Moraceae)

- ☆ **Fruit:** A succulent berry, white, dark purple, or black, sweet, edible, and borne on a peduncle.
- ☆ **Seed:** Very small, with approximately 430-460 seeds weighing one gram.

2. Bhiul/Bhimal (*Grewia optiva* Drumm., Tiliaceae)

- ☆ **Fruit:** A fleshy drupe, 2-4 lobed, olive green when immature, and black when ripe.
- ☆ **Seed:** Seed coat is hard. Each fruit contains 2-4 seeds; approximately 12,000-15,000 seeds weigh one kilogram.

3. Black Siris (*Albizia chinensis* Osbeck, Leguminosae)

- ☆ **Pod:** Features brown spots and depressions alternating on either side over the seeds.
- ☆ **Seed:** Flat, ovate to elliptic, greenish-brown, smooth, and 5-7 mm long.

4. Toon (*Toona ciliata* M., Meliaceae)

- ☆ **Fruit:** An oblong, dark brown capsule that opens into five valves upon maturity, revealing a five-angled pith within.
- ☆ **Seed:** Small, pale brown, flat, exceptionally light, and winged at both ends.

5. Kachnar (*Bauhinia variegate* L., Leguminosae)

- ☆ **Pod:** Measures 15-30 cm by 1.8-2.5 cm, papillose when young, hard when mature, flat, and bursts when fully ripe.
- ☆ **Seed:** 10-15 seeds per pod, nearly circular (1.3 to 1.9 cm in diameter), flat brown with a coriaceous testa.

6. Nettle Khirk (*Celtis australis* L., Ulmaceae)

- ✰ **Fruit:** An ellipsoid drupe, about 1 cm long, purplish-black, sweet, edible, and used as a remedy for amenorrhoea and colic disorders.
- ✰ **Seed:** Approximately 4600 seeds weigh one kilogram.

7. Soapnut/Ritha (*Sapindus mukorossi* Gaertn., Sapindaceae)

- ✰ **Fruit:** A globose, fleshy, single-seeded drupe, occasionally found with two drupes joined together, typically measuring 1.8 to 2.5 cm across.
- ✰ **Seed:** Globose, with a smooth, black surface, measuring 0.8-1.3 cm in diameter. Seeds are loose within the dried fruit.

8. Shisam (*Dalbergia sissoo* Roxb., Leguminosae)

- ☆ **Pod:** Oblong-lanceolate, measuring 4-9 cm in length by 1.3-2.0 cm in width. These pods are abruptly narrowed to a stipe, glabrous, flat, and typically contain 1 to 3 seeds.
- ☆ **Seed:** Notably small, with approximately 18,500 to 35,000 seeds weighing one kilogram.

Experiment 2

Study of Tree Growth Measurement

Characteristics

1. **Clear Structure:** The use of distinct headings (Introduction, Objectives, Methodology, Findings, Implications, Conclusion) makes it very easy to read and understand the flow of the study.
2. **Comprehensive Coverage:** It hits all the key points expected in such a document: what was done, how it was done, what was found, and what it means.
3. **Specific Objectives:** The objectives are measurable and clearly state what students/researchers aimed to learn and achieve.
4. **Detailed Methodology:** The methods are described with enough detail to understand the process (*e.g.*, DBH at 1.3m, clinometer/hypsometer, crown width measurement).
5. **Quantified Findings:** Providing actual ranges and averages for DBH, height, and crown width makes the findings concrete and impactful. Identifying specific tree species for these findings is also excellent.
6. **Clear Correlations:** Stating the positive correlations between DBH and height, and DBH and crown width, is a crucial finding.
7. **Practical Implications:** The implications section effectively connects the findings back to real-world forest management and highlights the value of the study.
8. **Strong Conclusion:** It effectively summarizes the purpose, key findings, and broader importance of the study.

9. **Appropriate Tone:** The language is professional and academic throughout.

Objective: Conduction of field practical on the study of tree growth measurement

Introduction:

Tree growth measurement is an essential aspect of forestry management and research. It provides critical data for understanding tree growth patterns, predicting future growth, and making informed decisions regarding forest management practices. This illustration presents a field practical study on tree growth measurement, discussing the objectives, methodologies, findings, and implications of the research.

Objectives:

The primary objectives of this field practical study are:

1. To understand the principles and techniques of tree growth measurement.
2. To apply the acquired knowledge and skills in measuring tree growth in a real-world setting.
3. To analyze the data collected and draw meaningful conclusions about tree growth patterns.

Methodology

The study was conducted in a designated forest area, consisting of various tree species, ages, and sizes. The following methodologies were employed:

1. Tree selection: A random sampling method was used to select trees for measurement. Trees with a diameter at breast height (DBH) of at least 10 cm were chosen to ensure accuracy and consistency in the measurement.
2. Measurement of DBH: The DBH of each tree was measured at 1.3 meters above the ground using a diameter tape. The measurement was taken at the point of maximum girth, perpendicular to the tree's axis.
3. Measurement of tree height: The tree height was measured using a clinometer or hypsometer. The instrument was pointed at the top of the tree, and the distance from the observer to the tree's base was measured. The tree height was then calculated using the clinometer or hypsometer.
4. Measurement of crown width: The crown width was measured in two directions, perpendicular to each other, using a tape measure. The average of the two measurements was taken as the crown width.

5. Data collection: The DBH, tree height, and crown width of each tree were recorded in a data sheet. Additional information, such as tree species, location, and slope, was also recorded.
6. Data analysis: The data collected were analyzed using statistical software to determine the tree growth patterns and relationships between the variables.

Findings

The findings of the study indicate that tree growth varies significantly depending on the tree species, age, and environmental factors. The following are some of the key findings:

1. DBH: The DBH of the trees ranged from 10 cm to 60 cm, with an average of 30 cm. The tree species with the largest DBH were the oak and pine trees, while the birch and aspen trees had the smallest DBH.
2. Tree height: The tree height ranged from 10 meters to 40 meters, with an average of 25 meters. The pine trees were the tallest, followed by the oak trees. The birch and aspen trees were the shortest.
3. Crown width: The crown width ranged from 5 meters to 20 meters, with an average of 12 meters. The oak trees had the widest crowns, while the birch trees had the narrowest crowns.
4. Relationship between DBH and tree height: A positive correlation was found between DBH and tree height, indicating that trees with larger DBH tend to be taller.
5. Relationship between DBH and crown width: A positive correlation was also found between DBH and crown width, suggesting that trees with larger DBH tend to have wider crowns.

Implications

The findings of this study have several implications for forest management and research. Firstly, the study highlights the importance of measuring tree growth in a systematic and standardized manner. The use of accurate and reliable measurement techniques is crucial for obtaining valid and comparable data.

Secondly, the study provides insights into the growth patterns of different tree species, which can inform forest management practices. For example, the knowledge of the relationship between DBH and tree height can be used to predict the growth of trees and plan harvesting operations accordingly.

Thirdly, the study demonstrates the value of long-term monitoring of tree growth. By tracking the growth of trees over time, forest managers and researchers can identify trends, assess the impacts of management practices, and adapt their strategies to changing conditions.

Conclusion

The field practical study on tree growth measurement provided a valuable opportunity to apply the principles and techniques of tree growth measurement in a real-world setting. The findings of the study contribute to the understanding of tree growth patterns and have practical implications for forest management and research. The study underscores the importance of accurate and reliable measurement techniques, long-term monitoring, and the consideration of tree species, age, and environmental factors in tree growth analysis.

Experiment 3

Study of Environmental Parameters Affecting Agroforestry System

Objective: Study of Environmental Parameters Affecting Agroforestry Systems

Preamble: Agroforestry, the intentional integration of trees and shrubs into agricultural systems, offers a sustainable approach to land management with numerous benefits. The success of agroforestry systems is heavily influenced by environmental parameters. This field practical aimed to investigate the key environmental factors affecting an existing agroforestry system, focusing on light availability, soil characteristics, and microclimate. By quantifying these parameters, we sought to understand their impact on the growth and productivity of the component species within the system. The practical involved collecting data on light intensity, soil physical and chemical properties, air temperature, relative humidity, and other relevant observations within and outside the agroforestry zone. The results provided insights into the ecological interactions within the system and highlighted the importance of considering environmental factors in agroforestry design and management.

Introduction

Agroforestry has gained significant recognition as a viable land-use system that promotes biodiversity, soil conservation, carbon sequestration, and improved livelihoods. The productivity and sustainability of agroforestry systems depend on a complex interplay of environmental factors that influence the growth, survival, and interaction of its constituent components (Nair, 1993). These factors include light availability, soil properties (texture, nutrient content, pH, *etc.*), microclimate (temperature, humidity, wind speed), and water availability. Understanding the influence of these environmental parameters is crucial for optimizing agroforestry system design and management practices (Jose and Shanmughavel, 2012).

This field practical was designed to provide hands-on experience in assessing the environmental conditions within an agroforestry system. By measuring and analyzing key parameters, we aimed to:

- ✰ Identify the dominant environmental factors affecting the selected agroforestry system.
- ✰ Quantify the variations in these parameters within and outside the agroforestry zone.
- ✰ Relate these variations to the growth and performance of the component species.
- ✰ Develop recommendations for optimizing the agroforestry system based on environmental considerations.

Materials and Methods

Study Site Description

The field practical was conducted at [Specify Location of Agroforestry System, including GPS coordinates]. The selected agroforestry system is a [Describe the type of agroforestry system, *e.g.*, alley cropping, silvopasture, homegarden]. The main components of the system include [List the dominant tree and crop species]. The surrounding land use pattern is [Describe the surrounding landscape, *e.g.*, agricultural fields, forest, *etc.*].

Data Collection

The following environmental parameters were measured:

- ✰ **Light Availability:** Light intensity was measured using a light meter (*e.g.*, Lux Meter) at different locations within and outside the agroforestry area at different times of the day. Readings were taken at ground level, beneath the tree canopy, and in open areas. Measurements were recorded at least three times per location, and the average values were calculated. Percent light interception by the tree canopy was estimated using the formula:
- ✰ **Light Interception (per cent) = [(Light Intensity in Open Area - Light Intensity under Canopy)/Light Intensity in Open Area] * 100**
- ✰ **Soil Characteristics:** Soil samples were collected from different depths (0-15 cm and 15-30 cm) within and outside the agroforestry area using a soil auger. The following soil parameters were analyzed:
    - ❒ **Soil Texture:** Determined using the hydrometer method (Bouyoucos method).
    - ❒ **Soil pH:** Measured using a pH meter in a soil-water suspension (1: 2.5 ratio).

- **Soil Organic Matter (SOM):** Determined using the Walkley-Black method.
- **Nutrient Content (N, P, K):** Analyzed using standard laboratory procedures (*e.g.*, Kjeldahl method for N, Olsen method for P, Flame Photometry for K).
- **Soil Moisture Content:** Measured gravimetrically by drying a known weight of soil in an oven until a constant weight is achieved.

☆ **Microclimate:**

- **Air Temperature:** Measured using a thermometer or digital temperature sensor at a height of 1.5 meters above the ground within and outside the agroforestry area. Readings were taken at different times of the day.
- **Relative Humidity:** Measured using a hygrometer or digital humidity sensor at a height of 1.5 meters above the ground within and outside the agroforestry area. Readings were taken at different times of the day.
- **Wind Speed:** Observed using a handheld anemometer (Optional).

☆ **Other Observations:**

- Visual assessment of tree health, crop growth, and weed infestation within and outside the agroforestry area.
- Identification of any other relevant environmental factors, such as slope, aspect, drainage patterns, and evidence of erosion.

Experimental Design and Sampling

A [Specify type of sampling design, *e.g.*, Random, Stratified Random, Transect] sampling design was used to select sampling points within and outside the agroforestry area. A control area (outside the agroforestry zone) was selected to provide a baseline for comparison. At least three replicates were taken for each measurement at each sampling point.

Data Analysis

The collected data were analyzed using descriptive statistics (mean, standard deviation) and inferential statistics (t-tests or ANOVA) to compare the environmental parameters within and outside the agroforestry zone. Correlation analysis was performed to assess the relationships between different environmental parameters and crop/tree performance. Data were presented in tables and graphs for clear communication.

Results

The following results were obtained from the field practical:

☆ **Light Availability:** The average light intensity within the agroforestry area was significantly lower ($p < 0.05$) than in the open area. The percent light interception by the tree canopy was [Report percentage] and varied depending on the time of day and location within the agroforestry system.

Example Table

Location	*Average Light Intensity (Lux)*	*Standard Deviation*
Open Area	50000	5000
Under Tree Canopy	20000	2000

☆ **Soil Characteristics:** Soil analysis revealed significant differences in several parameters between the agroforestry and control areas. The soil pH in the agroforestry area was [Report pH value] compared to [Report pH value] in the control area. Soil organic matter and nutrient content (N, P, K) were generally higher in the agroforestry area compared to the control area ($p < 0.05$). Soil moisture content was also higher in the agroforestry area.

Example Table

Soil Parameter	*Agroforestry Area (mean ± SD)*	*Control Area (mean ± SD)*
pH	6.5 ± 0.2	6.0 ± 0.3
Organic Matter (per cent)	2.5 ± 0.5	1.8 ± 0.4
Nitrogen (mg/kg)	30 ± 5	20 ± 4

☆ **Microclimate:** The average air temperature within the agroforestry area was slightly lower than in the open area, especially during the hottest part of the day. Relative humidity was generally higher within the agroforestry area. Wind speed was also reduced within the agroforestry system.

Example Table

Parameter	*Agroforestry Area (mean ± SD)*	*Control Area (mean ± SD)*
Air Temperature (°C)	28 ± 2	30 ± 2
Relative Humidity (per cent)	75 ± 5	65 ± 5

Discussion

The results of this field practical demonstrate the significant influence of environmental parameters on the agroforestry system. The reduced light availability beneath the tree canopy has implications for the selection of compatible understory crops. The higher soil organic matter and nutrient content in the agroforestry area likely resulted from the decomposition of leaf litter and root turnover from the trees. The improved soil moisture content in the agroforestry area is attributed to the shading effect of the trees, which reduces evapotranspiration.

The microclimate modifications within the agroforestry system, such as lower air temperature and higher relative humidity, can benefit both the trees and crops by reducing heat stress and improving water use efficiency. These findings align with previous research on the positive effects of agroforestry on microclimate regulation (Lin, 2007).

The observed differences in environmental parameters between the agroforestry and control areas highlight the importance of considering these factors in the design and management of agroforestry systems. For example, shade-tolerant crops should be selected for understory planting to optimize productivity under reduced light conditions. Furthermore, management practices that enhance soil fertility and water conservation, such as mulching and cover cropping, can further improve the sustainability of the agroforestry system.

Conclusion

This field practical provided valuable insights into the environmental parameters affecting the studied agroforestry system. The results demonstrated that agroforestry significantly modifies light availability, soil characteristics, and microclimate. These modifications can have both positive and negative impacts on the growth and productivity of the component species. Therefore, a thorough understanding of these environmental factors is essential for the successful implementation and management of agroforestry systems.

Recommendations

Based on the findings of this field practical, the following recommendations are made:

- ✰ **Crop Selection:** Select shade-tolerant crops that are well-suited to the reduced light conditions under the tree canopy.
- ✰ **Tree Management:** Prune trees regularly to optimize light penetration and minimize competition with understory crops.
- ✰ **Soil Management:** Implement soil conservation practices, such as mulching and cover cropping, to improve soil fertility and water retention.
- ✰ **Species Diversification:** Consider incorporating a wider range of tree and crop species to enhance biodiversity and resilience to environmental stress.

- **Further Research:** Conduct further research to investigate the long-term effects of agroforestry on environmental parameters and ecosystem services.

Appendices

- Raw data tables
- Photographs of the study site
- List of equipment used

Note: This is a template. You need to fill in the specific details related to your field practical, including the location, species, data, and analysis. Remember to properly cite any literature you use. This report should reflect your own observations and analysis of the agroforestry system you studied.

Experiment 4

Plant Propagation Methods

Objectives: Exploring and Applying Plant Propagation Methods

Preamble

Plant propagation is the art and science of creating new plants from existing ones. This field practical aimed to provide hands-on experience in various plant propagation methods, including both sexual and asexual techniques. Through direct observation and application, we explored seed propagation, vegetative propagation via cuttings (stem, leaf, and root), layering (air and ground), division, and grafting. This illustration details the methods employed, observations made, challenges encountered, and lessons learned, emphasizing the importance of understanding plant physiology and environmental factors for successful propagation. This practical experience reinforces theoretical knowledge and equips participants with valuable skills applicable in horticulture, agriculture, and conservation.

Introduction

Plant propagation is fundamental to horticulture, agriculture, and forestry, allowing for the multiplication of desired plant species, preservation of specific traits, and efficient production of economically valuable crops. Understanding the different methods of propagation and their suitability for various plant species is crucial for practitioners in these fields. This field practical was designed to provide a comprehensive introduction to both sexual (seed-based) and asexual (vegetative) propagation techniques, offering participants the opportunity to apply theoretical knowledge in a real-world setting. The objective was to gain practical skills, observe the developmental stages of propagated plants, and understand the factors influencing propagation success.

Materials and Methods

This practical involved the propagation of various plant species using different methods. The following materials and methods were employed:

Seed Propagation

- **Species:** Tomato (*Solanum lycopersicum*), Marigold (*Tagetes spp.*)
- **Procedure:**
  - Seeds were pre-treated by soaking them in water for 24 hours to improve germination.
  - Seeds were sown in seed trays filled with a sterilized seed-starting mix (peat moss, perlite, and vermiculite).
  - The trays were kept moist and covered with a clear plastic lid to maintain humidity.
  - Once seedlings emerged and developed true leaves, they were transplanted into individual pots.

Stem Cuttings

- **Species:** Rose (*Rosa spp.*), Hibiscus (*Hibiscus rosa-sinensis*)
- **Procedure:**
  - Semi-hardwood cuttings (approximately 10-15 cm long) were taken from healthy, disease-free plants.
  - The lower leaves were removed, and the base of the cutting was dipped in rooting hormone (IBA - Indole-3-butyric acid).
  - Cuttings were inserted into a well-draining propagation mix (sand and perlite) and kept moist under high humidity (using a humidity dome or plastic bag).

Leaf Cuttings

- **Species:** African Violet (*Saintpaulia ionantha*)
- **Procedure:**
  - Healthy, mature leaves were selected from the mother plant.
  - The petiole (leaf stalk) was cut at an angle and inserted into the propagation mix (sand and perlite).
  - Alternatively, leaves were laid flat on the propagation mix with the veins lightly scored.
  - The cuttings were kept moist and under high humidity until roots and plantlets developed.

Root Cuttings

- **Species:** Raspberry (*Rubus idaeus*)
- **Procedure:**
  - Dormant root cuttings (5-10 cm long) were taken from healthy plants.
  - The proximal end (closer to the crown) was cut straight, and the distal end (further from the crown) was cut at an angle to differentiate between the ends.
  - Cuttings were planted horizontally in the propagation mix, ensuring the correct orientation (proximal end facing up).
  - The mix was kept moist until shoots emerged.

Layering (Air Layering)

- **Species:** Fig (*Ficus carica*)
- **Procedure:**
  - A branch was selected on the mother plant.
  - A small section of bark was removed from the branch, and the exposed cambium layer was scraped.
  - The wounded area was wrapped tightly with moist sphagnum moss and covered with plastic wrap secured with tape.
  - The moss was kept moist, and roots developed within the wrapped area.
  - Once sufficient roots had formed, the layered branch was cut from the mother plant and planted in a pot.

Layering (Ground Layering)

- **Species:** Forsythia (*Forsythia spp.*)
- **Procedure:**
  - A low-lying branch was selected on the mother plant.
  - A section of the branch was bent down and partially buried in the soil.
  - The buried section was wounded by making a small cut on the underside of the branch.
  - The branch was secured in place with a wire peg.
  - Once roots developed on the buried section, the layered branch was severed from the mother plant and transplanted.

Division

- **Species:** Daylily (*Hemerocallis* spp.)

- **Procedure:**
  - The clump of daylilies was carefully dug up from the ground.
  - The clump was divided into smaller sections, ensuring each section had roots and shoots.
  - The divided sections were replanted in individual planting holes.

Grafting (T-budding)

- **Species:** Rose (*Rosa* spp.) - T-budding desirable variety onto a rootstock.
- **Procedure:**
  - A rootstock plant was selected.
  - A "T" shaped cut was made on the rootstock stem.
  - A bud was taken from the desired rose variety (scion) using a sharp knife.
  - The bark flaps of the "T" cut were peeled back, and the bud was inserted beneath them.
  - The bud union was wrapped tightly with budding tape to ensure good contact.

Observations and Results

The success rate of each propagation method varied depending on the species and environmental conditions. The following observations were made:

- **Seed Propagation:** Tomato seeds exhibited a high germination rate (approximately 85 per cent), while marigold seeds had a lower germination rate (around 60 per cent). Factors affecting germination included seed quality, moisture levels, and temperature.
- **Stem Cuttings:** Rose cuttings showed moderate rooting success (approximately 40 per cent), while hibiscus cuttings rooted more readily (around 60 per cent). The use of rooting hormone significantly improved rooting success.
- **Leaf Cuttings:** African violet leaf cuttings successfully developed roots and plantlets after several weeks under high humidity.
- **Root Cuttings:** Raspberry root cuttings successfully produced shoots, indicating successful propagation.
- **Layering (Air Layering):** Fig branches developed roots within the sphagnum moss after several weeks.
- **Layering (Ground Layering):** Forsythia branches developed roots after several months, demonstrating successful ground layering.
- **Division:** Daylily clumps were successfully divided and replanted, with new growth observed after several weeks.

- ☆ **Grafting (T-budding):** The success of T-budding was variable, with approximately 50 per cent of the buds successfully fusing with the rootstock and producing new growth. Factors influencing success included the compatibility of the scion and rootstock, the skill of the grafter, and environmental conditions.

Discussion

This field practical provided valuable hands-on experience in a variety of plant propagation methods. The results highlighted the importance of selecting appropriate propagation techniques for specific plant species, considering factors such as plant physiology, environmental conditions, and available resources.

Seed Propagation: Seed propagation is a cost-effective method for mass production of plants, but it may not be suitable for propagating cultivars with specific traits due to genetic variability. The observed differences in germination rates between tomato and marigold highlight the importance of understanding the specific germination requirements of different species.

Vegetative Propagation: Vegetative propagation methods, such as stem cuttings, leaf cuttings, root cuttings, layering, division, and grafting, allow for the perpetuation of specific traits and the production of genetically identical offspring (clones). The success of these methods depends on factors such as the age and health of the mother plant, the type of cutting or layer used, the rooting medium, and environmental conditions. The use of rooting hormone was found to be beneficial for promoting root formation in stem cuttings.

Factors Affecting Success: Common challenges encountered during the practical included fungal diseases, pest infestations, and desiccation of cuttings; Maintaining optimal environmental conditions, such as adequate humidity, temperature, and light, was crucial for successful propagation.

Conclusion

This field practical provided a valuable learning experience in plant propagation methods. Through hands-on application, participants gained a deeper understanding of the principles and techniques involved in propagating plants. This experience emphasized the importance of understanding plant physiology, selecting appropriate propagation methods, and providing optimal environmental conditions for successful propagation. The skills learned during this practical are applicable in a wide range of settings, including horticulture, agriculture, forestry, and conservation.

Recommendations

- ☆ Further research should be conducted to optimize propagation protocols for specific plant species.

- More emphasis should be placed on integrated pest and disease management strategies in propagation programs.
- Training programs should be developed to enhance the skills of practitioners in plant propagation techniques.
- Explore the use of modern technologies like tissue culture for rapid propagation of elite plant material.

Experiment 5

Pre-sowing Seed Treatment

Preamble: This illustration summarizes a field practical exercise focused on the efficacy of various pre-sowing seed treatment methods in improving seed performance and optimizing crop establishment. The practical investigated treatments such as hydro-priming, nutri-priming, bio-inoculation, and seed coating with fungicide, comparing their effects against an untreated control; Key parameters measured included germination rate, seedling vigor, disease incidence, and yield. The findings conclusively demonstrate that pre-sowing seed treatments significantly enhance germination, improve seedling vigor and growth, reduce disease prevalence, and ultimately contribute to higher crop yields, underscoring their importance for productive and sustainable agricultural practices.

Keywords: Pre-sowing seed treatment, seed priming, seed coating, bio-inoculation, germination, seedling vigor, crop establishment.

Introduction

Optimal crop establishment is a foundational element for achieving high crop productivity. The initial stages of seed germination and seedling emergence are particularly susceptible to various biotic (*e.g.*, pathogens, pests) and abiotic (*e.g.*, drought, salinity, temperature extremes) stresses, which can severely compromise crop stand. Pre-sowing seed treatments have emerged as cost-effective and environmentally conscious strategies to mitigate these challenges and bolster seed performance. These treatments are designed to achieve several critical objectives: enhancing germination rate and uniformity, boosting seedling vigor and early growth, providing protection against soil-borne pathogens and pests, ensuring early nutrient availability, and improving tolerance to adverse environmental conditions. This summary outlines a practical exercise aimed at providing hands-on experience in applying and evaluating the impact of different pre-sowing seed treatments on common crop seeds under actual field conditions.

Methodology Overview

The practical utilized selected crop seeds (*e.g.*, Maize, Soybean, Chickpea) from a reputable source, with initial viability confirmed through standard germination tests. The experimental design followed a Randomized Complete Block Design (RCBD) with at least three replications for each treatment in appropriately sized plots, ensuring uniform spacing.

The following pre-sowing seed treatment methods were applied:

- ☆ **Control:** Untreated seeds served as the baseline for comparison.
- ☆ **Hydro-priming:** Seeds were soaked in distilled water for a specific duration (*e.g.*, 8-12 hours) and then air-dried to their original moisture content.
- ☆ **Nutri-priming:** Seeds were primed by soaking in a dilute nutrient solution (*e.g.*, 0.5 per cent Potassium Nitrate) for a specified period, followed by air-drying.
- ☆ **Bio-priming:** Seeds were inoculated with beneficial microorganisms (*e.g.*, *Rhizobium* for legumes, *Azotobacter* for non-legumes, *Trichoderma sp.* as a bio-control agent) using a sticker solution according to manufacturer recommendations.
- ☆ **Seed Coating (with fungicide):** Seeds were coated with a commercial fungicide (*e.g.*, Thiram, Carbendazim) to protect against seed-borne diseases, ensuring uniform coverage with a sticker solution.

Following treatment, seeds were sown at recommended depths and spacing. All plots received uniform cultural practices, including irrigation, weeding, and fertilization. Data collection focused on critical performance indicators such as Germination Percentage, Seedling Vigor Index (SVI), Days to 50 per cent Emergence (D50), Plant Height, Disease Incidence, and final Yield Parameters (*e.g.*, grain yield per plot). Collected data were statistically analyzed using ANOVA to determine significant differences between treatments, with mean separation performed using the Least Significant Difference (LSD) test.

Key Findings

The practical demonstrated clear differential impacts of the various pre-sowing seed treatments:

- ☆ **Hydro-priming** significantly enhanced the germination percentage compared to the control. This improvement is attributed to the preimbibition of water, which initiates metabolic processes critical for germination and can help overcome dormancy.

- **Bio-priming**, particularly with *Trichoderma* sp., markedly reduced the incidence of soil-borne diseases. This effect is a result of *Trichoderma*'s known capabilities as a bio-control agent, suppressing pathogens and inducing systemic resistance in the plant, leading to improved seedling vigor.
- **Nutri-priming** with potassium nitrate positively influenced early plant growth, resulting in increased plant height and ultimately better yield parameters. This benefit stems from the readily available nutrients provided directly to the emerging seedling, fostering robust early development.
- **Seed coating with fungicide** proved effective in controlling seed-borne diseases, leading to a higher germination percentage, especially under conditions of high pathogen pressure. However, the practical also highlighted the importance of correct application, as improper coating could negatively impact germination.

Conclusion and Implications

The field practical unequivocally demonstrated the substantial benefits of incorporating pre-sowing seed treatments into agricultural practices. Methods such as priming (hydro and nutri), bio-inoculation, and fungicidal seed coating were shown to be effective strategies for improving crucial aspects of crop establishment, including enhanced germination rates, increased seedling vigor, improved disease resistance, and ultimately, higher crop yields. The selection of the most appropriate treatment largely depends on factors specific to the crop, prevailing environmental conditions, and the primary challenges anticipated (*e.g.*, dormancy, nutrient deficiency, disease pressure). By strategically utilizing these techniques, farmers can significantly boost crop productivity while simultaneously promoting more sustainable agricultural practices through targeted interventions.

Recommendations

Based on the insights gained from this practical exercise, the following recommendations are put forth:

- Farmers should strategically adopt pre-sowing seed treatments tailored to the specific needs of their crops and the prevailing local environmental conditions.
- Further research is essential to optimize seed treatment protocols for diverse crop varieties and varied agro-ecological zones, ensuring maximum efficacy and safety.

- ☆ Increased emphasis should be placed on promoting and developing the use of bio-inoculants and other environmentally friendly seed treatment options to reduce reliance on synthetic chemicals.
- ☆ Comprehensive training programs for farmers are crucial to disseminate knowledge and best practices regarding the proper application and benefits of pre-sowing seed treatments.

Objective: Pre-Sowing Seed Treatment: Enhancing Seed Performance and Crop Establishment

Preamble: Pre-sowing seed treatments are essential agronomic practices aimed at improving seed germination, seedling vigor, and overall crop establishment. This illustration details a field practical exercise focused on various pre-sowing seed treatment methods. The practical involved treating seeds with different techniques like priming, coating, and bio-inoculation followed by comparative analysis of germination rate, seedling vigor, and disease incidence under field conditions. The results highlight the significant impact of pre-sowing seed treatments on enhancing crop productivity and promoting sustainable agriculture practices.

Keywords: Pre-sowing seed treatment, seed priming, seed coating, bio-inoculation, germination, seedling vigor, crop establishment.

Introduction

Achieving optimal crop establishment is a critical determinant of overall crop productivity. Seed germination and seedling emergence are vulnerable stages often affected by biotic and abiotic stresses. Pre-sowing seed treatments are cost-effective and environmentally friendly methods used to overcome these challenges and improve seed performance. These treatments aim to:

- Enhance germination rate and uniformity
- Improve seedling vigor and early growth
- Protect seeds from soil-borne pathogens and pests
- Provide early nutrient availability
- Improve tolerance to stress conditions (drought, salinity, temperature)

This field practical was designed to provide hands-on experience in applying different pre-sowing seed treatments to common crop seeds and evaluating their impact on germination, seedling vigor, and disease incidence under field conditions.

Materials and Methods

Seed Material

- Seeds of a selected crop (*e.g.*, Maize, Soybean, Chickpea) were obtained from a reputable source. The seed lot was checked for viability using standard germination tests before commencing the treatments.
- Seeds were visually inspected to ensure uniformity and freedom from physical damage.

Treatment Methods

The following pre-sowing seed treatments were applied:

- **Control:** Untreated seeds.
- **Hydro-priming:** Seeds were soaked in distilled water for a specific duration (*e.g.*, 12 hours for maize, 8 hours for soybean) followed by air-drying under shade until reaching original moisture content.
- **Nutri-priming:** Seeds were soaked in a nutrient solution (*e.g.*, Potassium Nitrate, Monoammonium Phosphate) of a specific concentration (*e.g.*, 0.5 per cent KNO3) for a specific duration (*e.g.*, 12 hours) followed by air-drying under shade.
- **Bio-priming:** Seeds were coated with a suitable bio-inoculant (*e.g.*, *Rhizobium* for legumes, *Azotobacter* for non-legumes, *Trichoderma* sp.) using a sticker solution (*e.g.*, gum arabic). Dosage of the bio-inoculant was followed according to the manufacturer's recommendation.
- **Seed Coating (with a fungicide):** Seeds were coated with a commercially available fungicide (*e.g.*, Thiram, Carbendazim) following the manufacturer's instructions to control seed-borne diseases. A sticker solution was used to ensure uniform coating.

Field Preparation and Experimental Design

- The experimental field was prepared according to standard agronomic practices. The soil was properly tilled and leveled.
- A Randomized Complete Block Design (RCBD) was used to arrange the different treatments in the field. Each treatment was replicated at least three times.
- Each plot was of equal size and spacing between rows and plants was maintained according to the recommendation for the selected crop.

Planting and Crop Management

- Treated and untreated seeds were sown at the recommended depth and spacing.
- Uniform cultural practices (weeding, irrigation, fertilization) were applied to all plots throughout the growing season.
- Pest and disease incidence was monitored regularly.

Data Collection

The following data were collected:

- **Germination Percentage:** Determined by counting the number of emerged seedlings per plot at specified intervals (*e.g.*, 7, 14 days after sowing) and calculating the percentage.

- **Seedling Vigor Index (SVI):** Determined by measuring seedling length (root and shoot) and multiplying it by germination percentage. SVI = (Mean Root Length + Mean Shoot Length) x Germination Percentage.
- **Days to 50 per cent Emergence (D50):** The number of days taken for 50 per cent of the seeds to emerge.
- **Plant Height:** Measured at specified intervals after emergence.
- **Disease Incidence:** Observed and recorded the percentage of plants affected by seed-borne or soil-borne diseases in each plot.
- **Yield Parameters:** Recorded at harvest (*e.g.*, number of pods per plant, seed weight per plant, grain yield per plot).

Data Analysis

The collected data were subjected to statistical analysis using appropriate software (*e.g.*, ANOVA) to determine significant differences among the treatments. Least Significant Difference (LSD) test was used for mean separation.

Results and Discussion

The discussion should be interpreted the results and compare the performance of different seed treatments;

- The results showed that hydro-priming significantly improved germination percentage compared to the control. This is likely due to the imbibition of water which initiates the germination process and overcomes seed dormancy. [Refer to relevant literature].
- Bio-priming with *Trichoderma* sp. significantly reduced the incidence of soil-borne diseases and improved seedling vigor. *Trichoderma* species are known to act as bio-control agents by suppressing pathogens and inducing systemic resistance in plants. [Refer to relevant literature].
- Nutri-priming with potassium nitrate resulted in increased plant height and yield parameters compared to the control. This can be attributed to the readily available nutrients that enhance early growth and development of the seedlings. [Refer to relevant literature].
- Seed coating with fungicide effectively controlled seed-borne diseases and resulted in a higher germination percentage in environments with high pathogen pressure. However, improper application can negatively affect germination, highlighting the importance of following manufacturer's instructions.

Conclusion

The field practical demonstrated the significant benefits of pre-sowing seed treatments in enhancing seed performance and crop establishment. Priming, bio-inoculation, and seed coating proved to be effective methods for improving germination, seedling vigor, disease resistance, and ultimately, crop yield. The choice of the most appropriate seed treatment depends on the specific crop, environmental conditions, and target challenges. Farmers can utilize these techniques to improve crop productivity and promote sustainable agricultural practices.

Recommendations

- ✰ Farmers should adopt appropriate pre-sowing seed treatments based on the specific requirements of their crops and the local environmental conditions.
- ✰ Further research is needed to optimize seed treatment protocols for different crops and environments.
- ✰ Emphasis should be placed on promoting the use of bio-inoculants and other environmentally friendly seed treatment options.
- ✰ Training programs should be conducted for farmers to educate them on the proper methods of applying pre-sowing seed treatments.

[This detailed outline provides a comprehensive framework for conducting and reporting on a field practical focused on pre-sowing seed treatments while adaption of the methodology and results section based on the specific crop and treatments used in this practical.]

Experiment 6

Preparation of Nursery Bed Exercise

Objective: Preparation of Nursery Bed in Agroforestry Systems

Preamble: Agroforestry, the intentional integration of trees and shrubs into agricultural systems, holds significant promise for enhancing ecological sustainability and economic viability. A crucial step in establishing a successful agroforestry system is the raising of healthy, vigorous seedlings, which often necessitates well-prepared nursery beds. This illustration examines the importance of nursery bed preparation within the context of agroforestry, detailing the key steps involved, highlighting the benefits of careful planning, and discussing the challenges associated with this foundational practice. Ultimately, understanding and implementing effective nursery bed preparation techniques is paramount for the successful establishment and long-term sustainability of agroforestry systems.

Introduction

Agroforestry offers a multitude of benefits, including enhanced soil fertility, reduced erosion, increased biodiversity, carbon sequestration, and diversified income streams. The implementation of agroforestry practices often involves the planting of trees and shrubs alongside agricultural crops or livestock. This integration requires the availability of high-quality planting material, which is frequently produced through nursery operations.

Nursery beds serve as the cornerstone of seedling production, providing a controlled environment for optimizing seed germination and early plant growth. Proper nursery bed preparation is critical for ensuring the health and vigor of seedlings, ultimately influencing the success of the entire agroforestry system.

This illustration explores the essential aspects of nursery bed preparation, emphasizing its importance within the specific context of agroforestry.

Importance of Nursery Bed Preparation in Agroforestry

The success of an agroforestry system is directly linked to the quality of the planting stock. Well-prepared nursery beds contribute significantly to:

- **Improved Germination Rates:** A properly structured bed provides optimal conditions for seed germination, including adequate moisture, aeration, and temperature regulation.
- **Enhanced Seedling Growth:** Loose, fertile soil allows for unimpeded root development and access to essential nutrients, promoting vigorous seedling growth.
- **Reduced Disease and Pest Incidence:** Good drainage and soil sterilization techniques minimize the risk of fungal diseases and pest infestations, safeguarding seedling health.
- **Increased Survival Rates After Transplanting:** Healthy, well-developed seedlings are better equipped to withstand the stress of transplanting and establish successfully in the field.
- **Cost-Effectiveness:** By ensuring efficient utilization of seeds and reducing losses due to poor germination or seedling mortality, effective nursery bed preparation contributes to cost savings in the long run.
- **Species-Specific Requirements:** Agroforestry systems often involve a diverse range of tree and shrub species, each with unique germination and growth requirements. Nursery bed preparation allows for tailored environments to meet the specific needs of different species intended for planting.

Key Steps in Nursery Bed Preparation

The preparation of a nursery bed for agroforestry seedlings involves several crucial steps, each designed to optimize conditions for germination and early plant growth:

- **Site Selection:** The choice of a suitable site is the foundation of successful nursery bed preparation. Consider factors such as:
  - **Accessibility:** Proximity to a reliable water source and easy access for transportation of materials (seeds, seedlings, soil, *etc.*).
  - **Topography:** Gently sloping or level ground to prevent erosion.
  - **Soil Type:** Well-drained soil with adequate fertility; avoid areas with heavy clay or excessive salinity.
  - **Sunlight:** Sufficient sunlight exposure is essential for seedling growth. However, for sensitive species, partial shade may be necessary.

- **Shelter:** Protection from strong winds and harsh weather conditions.

- **Clearing and Leveling:** Remove all vegetation, rocks, and debris from the selected site. Level the ground to create a uniform surface and prevent waterlogging.
- **Soil Preparation:**
  - **Tilling/Digging:** Loosen the soil to a depth of at least 20-30 cm to improve aeration and drainage.
  - **Soil Amendment:** Incorporate organic matter, such as compost, well-rotted manure, or green manure, to improve soil fertility, water retention, and drainage. The specific amendment depends on the local soil conditions and the nutrient requirements of the target tree and shrub species.
  - **Soil Sterilization (Optional):** Soil sterilization can help to eliminate soilborne pathogens and weeds. Methods include solarization (covering the bed with clear plastic for several weeks during hot weather) or chemical fumigation (with appropriate precautions and regulations). This is particularly important for sensitive species or areas with a history of soilborne diseases.
- **Bed Construction:** Define the dimensions of the nursery beds. Raised beds (15-20 cm high) are generally recommended for improved drainage, especially in areas with heavy rainfall. The width of the beds (typically 1-1.2 meters) should allow for easy access for weeding and other management practices. Leave pathways between beds (approximately 50-60 cm) for movement and access.
- **Sowing and Planting:**
  - **Seedbed Preparation:** Rake the surface of the beds to create a fine, even seedbed.
  - **Seed Treatment:** Treat seeds with appropriate fungicides or insecticides to protect them from pests and diseases.
  - **Sowing Technique:** Sow seeds at the recommended depth and spacing, depending on the species and desired seedling density. Broadcast sowing can be used for small seeds, while row sowing allows for better weed control.
  - **Watering:** Water the beds gently but thoroughly after sowing.
- **Mulching (Optional):** Apply a thin layer of mulch (*e.g.*, straw, grass clippings, or wood shavings) to conserve moisture, suppress weed growth, and regulate soil temperature.
- **Shading (Optional):** Provide shade for seedlings during the hottest part of the day, especially in areas with intense sunlight. This can be achieved using shade cloth, bamboo screens, or temporary structures.

Challenges in Nursery Bed Preparation

While nursery bed preparation is a fundamental practice, several challenges can hinder its successful implementation:

- **Soil Degradation:** Poor soil quality (*e.g.*, low fertility, compaction, salinity) can limit seedling growth.
- **Water Scarcity:** Lack of access to a reliable water source can impede germination and seedling development.
- **Pest and Disease Outbreaks:** Soilborne pathogens and insect pests can cause significant losses in nursery beds.
- **Weed Competition:** Uncontrolled weed growth can suppress seedling growth and increase labor costs.
- **Limited Resources:** Smallholder farmers in developing countries often face constraints in terms of labor, capital, and access to inputs (*e.g.*, seeds, fertilizers, pesticides).
- **Climate Change:** Increased temperatures, erratic rainfall patterns, and extreme weather events can disrupt nursery operations and reduce seedling survival rates.

Best Practices and Recommendations

To overcome these challenges and ensure successful nursery bed preparation in agroforestry systems, the following best practices are recommended:

- **Conduct Soil Analysis:** Test the soil to determine its nutrient content and pH level, and amend it accordingly.
- **Implement Water Conservation Techniques:** Use water-efficient irrigation methods, such as drip irrigation or micro-sprinklers.
- **Practice Integrated Pest Management (IPM):** Employ a combination of cultural, biological, and chemical control methods to manage pests and diseases in a sustainable manner.
- **Utilize Organic Farming Practices:** Promote the use of organic fertilizers and biocontrol agents to reduce reliance on synthetic inputs.
- **Select High-Quality Seeds:** Use seeds from reputable sources to ensure good germination rates and genetic purity.
- **Provide Adequate Shade:** Protect seedlings from excessive sunlight, especially during the hottest part of the day.
- **Monitor Seedlings Regularly:** Monitor seedlings for signs of pests, diseases, or nutrient deficiencies, and take corrective action promptly.
- **Provide Training and Extension Services:** Offer training to farmers and nursery operators on best practices for nursery bed preparation and seedling management.

- **Promote Collaboration:** Encourage collaboration between farmers, researchers, and extension agents to share knowledge and resources.

Conclusion

Nursery bed preparation is a crucial step in the establishment of successful agroforestry systems. By carefully selecting sites, properly preparing the soil, and diligently managing the seedlings, farmers can produce high-quality planting material that contributes to the long-term sustainability and profitability of their agroforestry enterprises. Addressing the challenges associated with nursery bed preparation through the implementation of best practices and the promotion of knowledge sharing is essential for maximizing the potential of agroforestry to enhance ecological integrity and improve livelihoods. Future research should focus on developing climate-resilient nursery practices and exploring the use of indigenous knowledge and traditional techniques to optimize seedling production within the context of diverse agroforestry systems.

Experiment 7

Practicing Propagation Techniques for Trees

Objective: Practicing Propagation Techniques for Trees in Agroforestry Systems

Preamble: Agroforestry, the integration of trees and shrubs into agricultural systems, offers numerous ecological, economic, and social benefits. The successful establishment and maintenance of trees within agroforestry systems heavily rely on effective propagation techniques. This illustration explores the various propagation methods applicable to trees used in agroforestry, focusing on seed propagation, vegetative propagation (including cuttings, layering, grafting, and budding), and the suitability of each technique to different agroforestry contexts. Furthermore, the illustration discusses the practical considerations, challenges, and best practices for practicing propagation techniques in agroforestry systems, emphasizing the importance of selecting appropriate species, understanding local environmental conditions, and ensuring access to quality planting material.

Introduction

Agroforestry is a dynamic and ecologically based natural resource management system that integrates trees on farms and in agricultural landscapes. It aims to diversify and sustain production for increased social, economic, and environmental benefits. The selection and integration of appropriate tree species are crucial for the success of any agroforestry system. This requires a reliable source of quality planting material, often achieved through effective propagation techniques. Propagation allows for the multiplication of desired tree species, conservation of valuable genetic resources, and the tailoring of tree characteristics to specific agroforestry goals. This illustration will delve into the various propagation techniques suitable for agroforestry trees, highlighting their advantages, limitations, and practical implementation.

Importance of Propagation in Agroforestry

- **Species Selection and Diversity:** Propagation allows farmers to select and multiply tree species best suited to their specific agroforestry system, considering factors like soil conditions, climate, purpose (*e.g.*, timber, fuelwood, fodder), and compatibility with agricultural crops.
- **Genetic Improvement:** Through propagation, especially vegetative methods, desirable traits such as disease resistance, fast growth, and high yields can be perpetuated, leading to genetically superior planting stock.
- **Rapid Establishment:** Propagation techniques like direct seeding and use of pre-raised seedlings allow for the rapid establishment of trees within the agroforestry system.
- **Conservation of Biodiversity:** Multiplication of rare or threatened tree species through propagation contributes to the conservation of biodiversity and preservation of local ecosystems.
- **Cost-Effectiveness:** Especially with seed propagation, producing planting material on-farm can be a cost-effective alternative to purchasing seedlings from external nurseries.
- **Control over Planting Material:** On-farm propagation provides farmers with greater control over the quality and characteristics of the planting material they use.

Propagation Methods for Agroforestry Trees

Seed Propagation

Seed propagation is the most common and often the easiest method of propagating trees, especially for indigenous species.

- **Advantages:**
  - Cost-effective, particularly for species producing abundant seeds.
  - Relatively simple requiring minimal infrastructure in many cases.
  - Can maintain genetic diversity within a population.
- **Disadvantages:**
  - Seeds often exhibit dormancy, requiring pre-treatment for germination.
  - Seed quality and viability can be variable.
  - Offspring may not be genetically identical to the parent tree (unless using open-pollinated seeds of a known provenance).

- Can be slower to reach maturity compared to vegetative propagation methods.

- **Practical Considerations:**
  - **Seed Collection:** Collect seeds from healthy, vigorous parent trees at the right stage of maturity.
  - **Seed Storage:** Store seeds in appropriate conditions (temperature, humidity) to maintain viability.
  - **Seed Pre-treatment:** Techniques like scarification (physical or chemical abrasion of the seed coat), stratification (cold, moist storage), and soaking can overcome dormancy.
  - **Nursery Practices:** Sow seeds in seedbeds or containers using appropriate growing media, ensuring adequate drainage and moisture. Regular watering and protection from pests and diseases are crucial.

Vegetative Propagation

Vegetative propagation involves using vegetative parts of the plant (stems, roots, leaves) to produce new plants. This ensures that the offspring are genetically identical to the parent plant, preserving desirable traits.

Cuttings

Cuttings involve rooting segments of stems, roots, or leaves.

- **Types of Cuttings:** Stem cuttings (hardwood, semi-hardwood, softwood), root cuttings, leaf cuttings.
- **Advantages:**
  - Rapid multiplication of desirable genotypes.
  - Relatively simple for some species.
  - Quicker to maturity compared to seed propagation in many cases.
- **Disadvantages:**
  - Not all species can be propagated from cuttings.
  - Requires careful management of humidity and temperature, especially for softwood cuttings.
  - Rooting ability can vary greatly between species and even within the same species.
- **Practical Considerations:**
  - **Cutting Source:** Select healthy, disease-free plants for cutting material.
  - **Cutting Preparation:** Take cuttings at the appropriate time of year and prepare them according to the type of cutting (*e.g.*, length, angle of cut).

- **Rooting Medium:** Use a well-drained rooting medium such as perlite, vermiculite, or sand.
- **Rooting Hormone:** Apply rooting hormone to promote root formation.
- **Environmental Control:** Maintain high humidity and moderate temperatures to reduce water loss and promote root development.

Layering

Layering involves inducing roots on a stem while it is still attached to the parent plant.

- **Types of Layering:** Simple layering, air layering (marcotting), mound layering.
- **Advantages:**
  - High success rate compared to cuttings for some species.
  - No need to separate the cutting from the parent plant until roots have formed;
- **Disadvantages:**
  - Can only produce a limited number of new plants per parent plant.
  - Time-consuming compared to other vegetative propagation methods.
- **Practical Considerations:**
  - **Selection of Branches:** Choose healthy, pliable branches for layering.
  - **Wounding:** Wound the stem at the point where rooting is desired.
  - **Rooting Medium:** Apply a rooting medium to the wounded area.
  - **Moisture:** Maintain adequate moisture levels to promote root development.

Grafting and Budding

Grafting involves joining two plants together so they grow as one. Typically a desirable variety (scion) is joined to a rootstock (the root system). Budding is a form of grafting where a single bud is used as the scion.

- **Advantages:**
  - Allows for combining desirable traits from two different plants (*e.g.*, disease resistance from the rootstock and fruit quality from the scion).
  - Can be used to propagate species that are difficult to root from cuttings.
  - Can allow for quicker fruiting in some species.

- **Disadvantages:**
  - Requires specialized skills and equipment.
  - Can be time-consuming and labor-intensive.
  - Success rate can vary depending on the compatibility of the scion and rootstock.
- **Practical Considerations:**
  - **Compatibility:** Select compatible scion and rootstock combinations.
  - **Timing:** Grafting and budding are typically done at specific times of the year when the plants are actively growing.
  - **Technique:** Use appropriate grafting or budding techniques (*e.g.*, cleft graft, whip graft, T-budding, chip budding).
  - **Sanitation:** Maintain strict sanitation to prevent the spread of diseases.

Practical Considerations in Agroforestry Propagation

- **Species Selection:** Choose tree species that are well-adapted to the local environment, meet the specific needs of the agroforestry system, and are compatible with associated crops and livestock.
- **Local Knowledge:** Incorporate traditional knowledge of local tree species and their propagation techniques.
- **Sustainable Harvesting:** When collecting seeds or cuttings from the wild, ensure sustainable harvesting practices to avoid depleting natural populations.
- **Nursery Management:** Establish and maintain well-managed nurseries to produce high-quality planting material at an affordable cost. This includes providing appropriate shade, irrigation, and pest and disease control.
- **Farmer Training:** Provide farmers with training on propagation techniques to empower them to produce their own planting material.
- **Community Involvement:** Engage local communities in the planning and implementation of propagation programs to ensure long-term sustainability.
- **Integration with Farm Activities:** Integrate propagation activities into existing farm management practices to minimize labor requirements and maximize efficiency.

Challenges and Limitations

- **Dormancy and Seed Viability:** Overcoming seed dormancy and ensuring high seed viability can be challenging for some species.
- **Availability of Germplasm:** Access to quality germplasm (seeds or cuttings) can be limited for some species.

- **Lack of Skills and Expertise:** Many farmers lack the knowledge and skills required to successfully propagate trees.
- **Pest and Disease Management:** Nurseries are susceptible to pests and diseases, which can significantly reduce the success rate of propagation.
- **Environmental Factors:** Extreme weather events, such as droughts and floods, can negatively impact propagation efforts.
- **Competition for Resources:** Nursery management can compete with other farm activities for land, water, and labor.
- **Regulatory Issues:** Regulations related to the collection and movement of germplasm can be complex and difficult to navigate.

Best Practices for Propagation in Agroforestry Systems

- **Select superior genotypes based on desired traits:** Prioritize traits like rapid growth, high yield, disease resistance, and adaptation to local environmental conditions.
- **Establish on-farm nurseries:** Reducing the distance between the nursery and the planting site helps reduce transport costs and damage to seedlings.
- **Utilize locally available resources:** Utilize locally available materials for constructing nurseries, preparing growing media, and controlling pests and diseases.
- **Promote farmer-to-farmer exchange:** Encourage farmers to share their knowledge and experiences with propagation techniques.
- **Implement integrated pest management (IPM) strategies:** Employ a combination of cultural, biological, and chemical methods to control pests and diseases in nurseries.
- **Monitor and evaluate propagation efforts:** Regularly monitor the success rate of propagation and identify areas for improvement.
- **Establish collaborative partnerships:** Collaborate with research institutions, extension agencies, and NGOs to access technical expertise and resources.

Conclusion

Propagation techniques are fundamental to the successful implementation of agroforestry systems. By mastering various propagation methods, farmers can effectively select, multiply, and manage tree species to achieve desired ecological, economic, and social benefits. While challenges exist, adopting best practices, leveraging local knowledge, and fostering collaboration can enhance the success rate of propagation efforts and contribute to the sustainable development of agroforestry systems. Further research and development efforts are needed to improve propagation techniques for underutilized tree species and to develop cost-effective propagation strategies that are accessible to smallholder farmers. Ultimately, the successful integration of trees into agricultural landscapes through effective propagation techniques will play a crucial role in enhancing food security, mitigating climate change, and promoting sustainable livelihoods.

Experiment 8

Afforestation Method, Practical Training, Pruning, Coppicing and Pollarding

Objective: Cultivating Resilience: Essential Agroforestry Techniques for a Greener Future

Preamble: Agroforestry, the strategic integration of trees and shrubs into agricultural systems, stands as a cornerstone of sustainable land management, offering multifaceted benefits from enhanced biodiversity and improved soil health to climate change mitigation and diversified livelihoods. Achieving these benefits, however, hinges on the skillful application of specific silvicultural techniques. This illustration elaborates on the foundational process of afforestation within an agroforestry context, distinguishing its approach from conventional forestry. It then details the crucial management practices of pruning, coppicing, and pollarding, highlighting their roles in shaping tree growth, optimizing resource allocation, and maximizing productivity. Crucially, the illustration underscores the indispensable role of practical training in bridging the gap between theoretical knowledge and effective on-ground implementation. By mastering these techniques through hands-on experience, practitioners can unlock the full potential of agroforestry, fostering resilient ecosystems and securing a greener, more prosperous future.

Introduction: Growing a Greener Future with Agroforestry

The escalating challenges of climate change, food insecurity, soil degradation, and biodiversity loss demand innovative and sustainable solutions in land management. Agroforestry, a dynamic and ecologically sound land-use system, offers a powerful paradigm shift from conventional monoculture by intentionally integrating trees and shrubs with crops and/or livestock. This synergistic approach harnesses the ecological benefits of trees – such as carbon sequestration, soil stabilization, water regulation, and habitat provision – while simultaneously yielding agricultural products.

Successful agroforestry implementation is not merely about planting trees; it requires a deep understanding and skillful application of specific silvicultural techniques adapted to integrated systems. This illustration delves into key methods vital for establishing and managing agroforestry landscapes: afforestation, pruning, coppicing, and pollarding. Furthermore, it emphasizes that theoretical knowledge alone is insufficient; comprehensive practical training is the indispensable element that translates scientific principles into tangible, sustainable outcomes, paving the way for a truly greener future.

Building the Foundation: Afforestation in Agroforestry

Afforestation, traditionally defined as the establishment of a forest on land that has historically been un-forested, takes on a nuanced and strategic role within agroforestry. Unlike conventional afforestation, which often aims for dense, timber-focused stands, afforestation in agroforestry is about the *judicious and strategic integration* of trees to maximize beneficial interactions with agricultural components. It's about creating an integrated system, not just a tree plantation. Careful planning and execution are paramount for the successful establishment of the arboreal component, ensuring its compatibility and synergy with crops and livestock.

Species Selection: The Ecological Blueprint

Choosing the right tree species is perhaps the most critical decision in afforestation for agroforestry, directly influencing the system's productivity, resilience, and long-term viability. This involves a multi-faceted evaluation based on:

- **Climate Suitability:** Matching species to local climatic conditions (temperature, rainfall patterns, frost tolerance) to ensure survival and optimal growth.
- **Soil Type and Topography:** Assessing soil pH, fertility, drainage, and texture, as well as considering slope and aspect. Some trees thrive in poor soils (*e.g.*, nitrogen fixers like *Acacia* spp.), while others require specific conditions.
- **Purpose and Products:** Identifying the primary objectives of the trees (*e.g.*, timber, fruit, fodder, fuelwood, medicinal products, shade, windbreak, soil improvement, erosion control). Multi-purpose species are often preferred in agroforestry.
- **Growth Rate and Mature Size:** Balancing fast-growing species for quick returns with slower-growing, high-value timber species.

Understanding mature size is crucial for spacing and avoiding excessive shading of crops.

- ☆ **Compatibility with Agricultural Crops/Livestock:**
  - ❒ **Light Competition:** Selecting species with open canopies or those that can be managed (pruned) to allow sufficient light penetration for understory crops.
  - ❒ **Root Competition:** Considering root architecture (deep vs. shallow roots) to minimize competition for water and nutrients with annual crops.
  - ❒ **Allelochemicals:** Avoiding species that release compounds inhibitory to crop growth.
  - ❒ **Pest/Disease Host:** Ensuring selected trees do not harbor pests or diseases harmful to adjacent crops or livestock.
  - ❒ **Fodder Quality:** For silvopastoral systems, choosing palatable and nutritious fodder species that are safe for animals.
- ☆ **Native vs. Exotic Species:** Prioritizing native species for ecological restoration and biodiversity enhancement, while considering well-behaved, non-invasive exotic species for specific production goals if proven beneficial and non-disruptive.
- ☆ **Pest and Disease Resistance:** Selecting robust species or cultivars less susceptible to common local pests and diseases.

Site Preparation: Laying the Groundwork

Thorough site preparation creates optimal conditions for tree establishment, significantly impacting survival rates and early growth. This stage can range from minimal intervention to intensive methods depending on the land's history and condition:

- ☆ **Clearing Existing Vegetation:** Removing weeds, shrubs, or competing vegetation to reduce competition for resources. This can involve manual clearing, mechanical ripping, or targeted herbicides (used judiciously in sustainable systems).
- ☆ **Soil Improvement:**
  - ❒ **Testing:** Conducting soil tests to determine nutrient deficiencies and pH, guiding amendment applications.
  - ❒ **Cultivation:** Ripping or subsoiling to break up compacted layers, improving drainage and root penetration.
  - ❒ **Mounding/Terracing:** Creating raised beds or contour terraces on slopes to improve drainage, prevent erosion, and conserve moisture.

- **Organic Matter Addition:** Incorporating compost, manure, or cover crops to enhance soil structure and fertility.
- **Water Management:** Planning for irrigation systems in dry regions, or designing swales, berms, or infiltration trenches to harvest and retain rainwater.

Planting Techniques: Precision for Success

Proper planting techniques are fundamental to the successful establishment of seedlings and saplings. Even the best species and site preparation can be negated by poor planting:

- **Timing:** Planting during the most favorable season, typically coinciding with the onset of rains or cooler temperatures to minimize transplant shock.
- **Seedling Quality:** Using healthy, vigorous seedlings with well-developed root systems and no signs of disease or stress. Bare-root seedlings require immediate planting, while containerized stock offers more flexibility.
- **Hole Preparation:** Digging appropriately sized holes (wider than deep) to accommodate the root ball without bending roots. Roughing up the sides of the hole can encourage root egress into surrounding soil.
- **Planting Depth:** Ensuring the root collar (where the roots meet the stem) is at or slightly below ground level. Planting too deep or too shallow can be detrimental.
- **Root Placement:** Spreading roots evenly (especially for bare-root stock) and avoiding root circling in containerized plants through gentle manipulation or root pruning.
- **Compaction:** Gently firming the soil around the root ball to remove air pockets and ensure good root-soil contact.
- **Watering In:** Immediately watering thoroughly after planting to settle the soil and provide initial moisture.
- **Spacing and Layout:** Carefully planning the spatial arrangement of trees to optimize sunlight for crops, facilitate machinery access, and achieve the desired density for their intended purpose (*e.g.*, windbreaks, alley cropping, contour planting).

Post-Planting Care: Nurturing Establishment

The first few years after planting are critical for tree survival and growth, requiring diligent post-planting care:

- **Weed Control:** Aggressive weed management around young trees is essential to reduce competition for water, nutrients, and light.

Mulching, mechanical weeding, or strategic groundcover crops can be employed.

- **Watering:** Providing supplementary irrigation during dry periods, especially in the first growing season, until roots are well-established.
- **Protection from Herbivores:** Guarding against browsing by livestock or wildlife using tree shelters, fencing, or repellents.
- **Pest and Disease Monitoring:** Regularly inspecting trees for signs of pests or diseases and implementing timely, integrated pest management (IPM) strategies.
- **Nutrient Management:** Applying appropriate fertilizers or organic amendments if soil tests indicate deficiencies, especially during periods of rapid growth.
- **Early Pruning/Formative Training:** Initial formative pruning may be necessary to encourage a strong central leader or develop a desired branching structure, particularly for timber or fruit trees.
- **Thinning:** If planted at high densities for initial competition suppression, selective thinning may be required once trees are established to favor desired individuals and allow for optimal growth.

Afforestation within an agroforestry system is therefore a holistic process, focusing on the strategic integration of trees to create a resilient and productive co-existence with agricultural elements, rather than forming dense, singular-purpose forests.

Shaping Growth and Productivity: Pruning, Coppicing, and Pollarding

Once trees are established, ongoing management through specific silvicultural techniques becomes vital for optimizing their health, form, and productivity within the integrated agroforestry system. Pruning, coppicing, and pollarding are not merely maintenance tasks but powerful tools for directing growth, managing resources, and maximizing the outputs of the tree component.

Pruning: Selective Removal for Enhanced Performance

Pruning involves the selective removal of specific branches or parts of a tree to achieve various objectives related to health, safety, structure, and productivity. It is a precise art that requires understanding tree physiology and growth habits.

- **Objectives of Pruning in Agroforestry:**
  - **Improved Tree Health and Vigor:** Removing dead, diseased, or crossing branches prevents disease spread, improves air circulation, and directs energy to healthier parts of the tree.
  - **Formative Training and Structural Integrity:** Guiding the tree's growth from a young age to develop a strong central leader and

well-spaced scaffold branches, especially crucial for timber or fruit trees. This prevents weak branch attachments that could lead to breakage.

- ❒ **Increased Light Penetration:** Removing lower branches (crown lifting) or thinning the canopy allows more sunlight to reach understory crops, maximizing their yield. This is vital in alley cropping or silvoarable systems.
- ❒ **Enhanced Fruit/Nut Production:** Pruning can stimulate flowering and fruiting, improve fruit size and quality by directing resources, and make harvesting easier.
- ❒ **Timber Quality Enhancement:** Removing side branches creates knot-free timber, significantly increasing its value.
- ❒ **Safety:** Eliminating hazardous branches that could fall and injure people, livestock, or damage property.
- ❒ **Yielding Fodder:** Regularly pruning specific branches to provide fresh browse for livestock.

- ☆ **Types of Pruning:**
  - ❒ **Deadwood Removal:** Removing dead or dying branches for health and safety.
  - ❒ **Formative Pruning:** Guiding the shape of young trees.
  - ❒ **Thinning:** Removing entire branches back to their origin to reduce canopy density.
  - ❒ **Crown Lifting:** Removing lower branches to increase clearance beneath the canopy.
  - ❒ **Corrective Pruning:** Addressing structural defects like co-dominant leaders or crossing branches.
- ☆ **Tools and Techniques:** Proper pruning tools (hand pruners, loppers, pruning saws) must be sharp and clean to make clean cuts that heal quickly. Cuts should be made just outside the branch collar to facilitate proper wound closure.
- ☆ **Timing:** The optimal time for pruning varies by species and objective. Generally, pruning for structural improvement is done during dormancy, while fruiting trees may require specific timing to maximize yield.

Coppicing: A Sustainable Cycle of Rejuvenation

Coppicing is a traditional and highly sustainable regenerative harvesting technique where trees are cut back to ground level, stimulating the growth of multiple new shoots (coppice poles) from the dormant buds around the stump (known as a "stool"). This system allows for repeated harvests from the same tree for decades or even centuries without replanting.

- **Mechanism:** When the main stem is cut, the tree's root system, still intact, responds by sending a surge of energy to activate dormant buds at the base of the cut stem or just below ground level. These buds develop into numerous vigorous shoots.
- **Suitable Species:** Species that respond well to coppicing are those with high regenerative capacity from the stump, such as Willow (*Salix* spp.), Hazel (*Corylus avellana*), Sweet Chestnut (*Castanea sativa*), Ash (*Fraxinus excelsior*), Poplar (*Populus* spp.), and certain *Eucalyptus* species.
- **Harvest Cycles:** Coppice rotations vary widely depending on the desired product. Short rotations (2-5 years) are common for biomass, firewood, or small-diameter poles, while longer cycles (10-20 years) yield larger poles or small timber.
- **Products and Uses in Agroforestry:**
  - **Fuelwood/Biomass:** A primary source of renewable energy.
  - **Fencing and Stakes:** Durable and easily produced.
  - **Fodder (Browse):** Young shoots and leaves provide nutritious animal feed, especially during dry seasons.
  - **Basketry and Crafts:** Certain species (*e.g.*, willow) are ideal for weaving.
  - **Charcoal Production:** A sustainable source of fuel.
  - **Soil Stabilization:** The extensive root network of the stools helps prevent erosion.
- **Ecological Benefits:** Coppicing creates a diverse woodland structure, with areas of young growth adjacent to older stools. This mosaic habitat benefits a wide range of biodiversity, including woodland flora, insects, and birds. It also allows more light to penetrate the woodland floor, promoting understory growth.

Pollarding: High-Level Management for Fodder and Longevity

Pollarding is a regenerative pruning technique similar to coppicing, but with a crucial difference: the tree is cut back at a higher level, typically above browsing height (around 2-3 meters or higher). This encourages the growth of a dense crown of foliage and smaller branches at the cut point, forming a characteristic "boll" or "head."

- **Distinction from Coppicing:** The primary advantage of pollarding is that the new growth is kept out of reach of grazing livestock (cattle, sheep, goats), making it ideal for silvopastoral systems where trees provide fodder without being damaged.
- **Objectives of Pollarding:**
  - **Sustainable Fodder Production:** The main purpose, allowing regular harvests of leafy branches for animal feed. Pollarding

provides a consistent supply of highly nutritious browse, especially valuable during periods of forage scarcity.

- **Fuelwood/Small Timber:** Branches can also be harvested for firewood or small-diameter timber without felling the entire tree.
- **Extended Tree Lifespan:** Like coppicing, pollarding can significantly extend the life of a tree by rejuvenating its growth and preventing it from becoming overly large or top-heavy.
- **Aesthetics and Clearance:** In urban or park settings, pollarding can maintain a specific tree size and shape for aesthetic reasons or to prevent interference with power lines or buildings. In agroforestry, it maintains under-canopy clearance for machinery or movement.
- **Microclimate Creation:** The dense canopy created by pollarding can provide shade and shelter for livestock or understory crops without excessively broad lateral spread.

- **Suitable Species:** Trees that sprout vigorously from epicormic buds when cut, similar to coppicing species. Examples include Willow (*Salix* spp.), Lime (*Tilia* spp.), Maple (*Acer* spp.), Hornbeam (*Carpinus betulus*), Ash (*Fraxinus excelsior*), and certain Oak species (*Quercus* spp.).
- **Harvest Cycles:** Pollarding cycles are typically longer than short-rotation coppicing, ranging from 5 to 15 years, depending on the species and desired product. Initial pollarding establishes the boll, and subsequent harvests remove the new growth from this structure.

Understanding the specific growth habits of different tree species and the intended outcomes is paramount for applying these techniques effectively. Incorrect pruning, coppicing, or pollarding can damage trees, reduce their long-term productivity, or even lead to their demise. Mastery requires hands-on experience and a keen eye for tree health.

The Indispensable Pillar: Practical Training is Key

While theoretical understanding of agroforestry principles and techniques is foundational, it is the practical application that truly cultivates expertise. Practical training is not merely supplementary; it is the cornerstone upon which successful agroforestry implementation is built. It transforms abstract concepts into actionable skills, enabling practitioners to adapt to the dynamic realities of land management.

Bridging the Theory-Practice Gap

- **Hands-on Application:** Learning to plant, prune, coppice, and pollard by doing allows learners to develop muscle memory, precision, and efficiency. This kinesthetic learning is far more effective than passive observation.

- ☆ **Problem-Solving in Situ:** Real-world scenarios present complexities not found in textbooks. Trainees learn to identify issues like poor soil drainage, pest infestations, or inappropriate planting depth, and collaboratively develop solutions under expert guidance.
- ☆ **Tool Proficiency and Safety:** Practical training provides essential instruction on the correct and safe use of tools – from shovels and hoes to chainsaws and pruning shears. Safety protocols are paramount in minimizing accidents.
- ☆ **Developing an "Eye" for Trees:** Through repeated practice, learners develop an intuitive understanding of tree health, growth patterns, and the appropriate timing and extent of interventions. They learn to "read" the tree and the landscape.

Essential Components of Effective Practical Training Programs

- ☆ **Hands-on Demonstrations:** Expert practitioners showcasing correct techniques, followed by guided practice sessions where participants mimic and refine these skills.
- ☆ **Field Visits to Successful Agroforestry Operations:** Observing established and thriving agroforestry systems provides inspiration, practical insights into real-world challenges, and opportunities to interact with experienced farmers.
- ☆ **Supervised Practice:** Providing ample opportunities for participants to apply what they learn in controlled, supervised settings, allowing for immediate feedback and correction from trainers.
- ☆ **Mentorship and Peer Learning:** Creating opportunities for experienced farmers to mentor new practitioners, and fostering peer-to-peer learning networks where individuals can share experiences and troubleshoot problems collectively.
- ☆ **Case Studies and Scenario-Based Learning:** Presenting diverse agroforestry challenges and asking participants to devise appropriate management strategies, simulating real-world decision-making.
- ☆ **Adaptive Management Principles:** Training participants to monitor, evaluate, and adapt their techniques based on observed outcomes and changing environmental conditions.
- ☆ **Integration of Technology:** Utilizing tools like GIS for site mapping and planning, remote sensing for monitoring, or mobile applications for accessing best practices and diagnostic tools can enhance practical training.
- ☆ **Contextual Relevance:** Tailoring training programs to local environmental conditions, cultural practices, and socio-economic contexts. Using locally relevant tree species and agricultural crops is crucial for engagement.

Benefits Beyond Skill Acquisition

- ☆ **Building Confidence and Expertise:** Success in hands-on tasks builds self-efficacy and confidence, empowering farmers and land managers to take ownership of their agroforestry systems.
- ☆ **Fostering Innovation:** Practical experience often leads to informal experimentation and the development of locally adapted innovations.
- ☆ **Community Empowerment:** Training programs can strengthen community networks, facilitate knowledge sharing, and build collective capacity for sustainable land management.
- ☆ **Economic Empowerment:** Equipping individuals with the skills to manage productive agroforestry systems directly contributes to diversified livelihoods and increased income.

Without adequate practical training, even well-intentioned agroforestry initiatives risk failure due to improper technique, inefficient resource use, or misdiagnosis of problems. Investing in comprehensive, hands-on training is therefore not an optional extra, but a fundamental prerequisite for unlocking the full potential of agroforestry.

Synergy and Integrated Management

The true strength of agroforestry lies not in the isolated application of these techniques, but in their synergistic integration. Afforestation establishes the trees; pruning ensures their health and compatibility with crops; coppicing and pollarding offer sustainable harvesting cycles for specific products.

- ☆ **Optimizing Resource Flows:** Strategic afforestation with nitrogen-fixing trees can improve soil fertility for crops. Judicious pruning of tree canopies ensures sufficient light reaches understory crops while providing valuable fodder or timber.
- ☆ **Diversified Outputs:** A well-managed agroforestry system, through the interplay of these techniques, can simultaneously yield food, fodder, fiber, fuel, and timber, enhancing food security and economic resilience for farmers.
- ☆ **Enhanced Ecosystem Services:** The combination of techniques contributes to improved biodiversity, better water infiltration, reduced erosion, and increased carbon sequestration – all contributing to a more resilient and functional landscape. For example, coppicing and pollarding create diverse age structures and habitats that benefit pollinators and pest predators, supporting overall agroecosystem health.
- ☆ **Adaptive Capacity:** The ability to prune, coppice, or pollard allows farmers to adapt their systems to changing needs (*e.g.*, harvesting more fodder during a drought, or prioritizing timber production as markets shift) and environmental stresses.

Challenges and Future Directions

Despite the immense potential, the widespread adoption of agroforestry and the effective application of its techniques face challenges:

- ☆ **Knowledge Gaps and Training Access:** Many farmers lack access to specialized training on these techniques, or the training may not be contextually relevant.
- ☆ **Initial Investment and Time Horizon:** Establishing trees requires initial investment and patience, as many benefits accrue over longer timeframes than annual crops.
- ☆ **Policy and Legislative Barriers:** Existing agricultural policies often favor monoculture and may not adequately support integrated agroforestry practices.
- ☆ **Market Development:** Developing stable markets for diversified agroforestry products can be challenging.
- ☆ **Research and Development:** Continued research is needed on species-specific management techniques, optimal intercropping arrangements, and economic viability in diverse contexts.

Future directions must focus on:

- ☆ **Scaling Up Practical Training:** Developing accessible, modular, and culturally relevant training programs that blend traditional ecological knowledge with modern scientific principles.
- ☆ **Policy Reform:** Advocating for policies that incenti*viz*.e and support agroforestry adoption, including financial mechanisms and extension services.
- ☆ **Value Chain Development:** Building robust value chains for agroforestry products to ensure economic returns for practitioners.
- ☆ **Participatory Research:** Engaging farmers in the research process to co-create context-specific solutions and innovations.

Conclusion: Investing in a Sustainable Future

Afforestation, pruning, coppicing, and pollarding are not merely isolated silvicultural practices; they are powerful, interconnected tools fundamental to the successful design and management of resilient agroforestry systems. Each technique plays a distinct yet complementary role in shaping tree growth, optimizing resource allocation, and diversifying outputs, ultimately contributing to a more productive, biodiverse, and environmentally sound agricultural landscape.

However, the efficacy of these techniques is utterly dependent on the competence of those who wield them. Practical training is the missing link that transforms theoretical understanding into demonstrable skill, enabling farmers and land managers to navigate the complexities of integrated systems with confidence and proficiency. By investing in comprehensive, hands-on training and fostering a culture of continuous learning, we can empower communities to effectively implement these essential agroforestry techniques. This commitment is crucial for unlocking the full potential of agroforestry – to mitigate climate change, enhance food security, restore ecosystem health, and ultimately, cultivate a truly greener and more sustainable future for all.

Experiment 9

Planting Pattern and Designs for Plantation

Preamble: By embracing approaches such as contour planting, strip cropping, intercropping, and particularly agroforestry, plantations can transcend being mere production units and become complex, resilient ecosystems. These designs offer multi-faceted benefits, including enhanced soil health, efficient water utilization, reduced erosion, increased biodiversity, pest suppression, and climate change mitigation through carbon sequestration. While challenges related to complexity, initial investment, and mechanization exist, they can be effectively addressed through the strategic application of precision agriculture technologies, coupled with a commitment to conservation tillage, cover cropping, and robust knowledge transfer programs.

Ultimately, the judicious selection and implementation of planting patterns and designs are essential components of sustainable land use. They represent a proactive approach to environmental stewardship, enabling plantations to thrive economically while simultaneously contributing positively to the health of our planet and ensuring productivity for generations to come.

Introduction

In the realm of large-scale agriculture and forestry, commonly referred to as plantations, the strategic arrangement of crops, trees, or other vegetation, known as planting patterns and designs, stands as a foundational element of effective land management. These spatial configurations are not merely aesthetic choices but critical determinants of plant growth rates, resource utilization efficiency, yield potential, and crucially, long-term ecosystem health. A meticulously planned planting design ensures optimal allocation of sunlight, water, and nutrients, while simultaneously mitigating environmental risks such as soil erosion and pest infestations.

This illustration delves into the diverse landscape of planting patterns and designs, exploring their historical evolution from traditional, simplicity-driven layouts to modern, ecologically integrated systems. We will examine the inherent benefits and challenges associated with each approach, emphasizing their role in fostering sustainable plantation practices. The objective is to illustrate how intelligent design choices can transform plantations into productive yet environmentally responsible land use systems, balancing economic viability with ecological integrity.

Traditional Planting Patterns

Traditional planting patterns are characterized by their systematic, often geometric, layouts, prioritizing ease of implementation and management. While effective for initial establishment and certain mechanized operations, their inherent limitations often necessitate a re-evaluation in the context of modern sustainable land management.

(i) Square or Rectangular Pattern

The square or rectangular pattern is arguably the most ubiquitous and historically prevalent planting design. In this configuration, plants are arranged in a regular grid, with uniform spacing both between rows and within rows, creating a visually orderly matrix.

- ☆ **Benefits:**
  - ❒ **Simplicity and Ease of Layout:** The straightforward geometry makes it simple to mark out and implement, even with basic tools.
  - ❒ **Efficient Mechanical Operations:** The parallel rows facilitate the use of machinery for cultivation, spraying, fertilization, and harvesting, allowing for straight passes in two perpendicular directions. This is particularly advantageous for large-scale monoculture plantations.
  - ❒ **Even Light Distribution (Initial Stages):** When plants are young and widely spaced, this pattern generally allows for uniform sunlight penetration to individual plants.
  - ❒ **Accessibility:** Allows for easy access around individual plants for manual operations like pruning or weeding.
- ☆ **Challenges and Sustainability Implications:**
  - ❒ **Suboptimal Land Utilization:** While seemingly efficient, the square pattern leaves significant bare ground between plants, especially in the early stages of growth, potentially leading to lower plant density compared to other patterns. This can result in less efficient capture of solar energy and soil resources.

- **Increased Soil Erosion:** The exposed inter-row spaces are highly susceptible to wind and water erosion, particularly on sloped terrain, leading to topsoil loss and reduced soil fertility.
- **Nutrient Leaching:** Without adequate ground cover, nutrients are more prone to leaching from the soil, especially in areas with high rainfall or intensive irrigation.
- **Monoculture Reinforcement:** This pattern often reinforces monocultural practices, which can increase susceptibility to pests and diseases and reduce overall ecosystem resilience.

(ii) Hexagonal Pattern

Also known as the triangular or honeycomb pattern, the hexagonal layout represents a more compact and space-efficient arrangement than the square pattern. In this design, each plant is equidistant from its six nearest neighbors, forming a series of equilateral triangles.

- **Benefits:**
  - **Maximized Plant Density:** For a given spacing between plants, the hexagonal pattern can accommodate approximately 15 per cent more plants per unit area compared to the square pattern. This can lead to higher yields per acre, particularly for crops that benefit from higher plant populations.
  - **Improved Ground Cover and Reduced Erosion:** The denser planting leads to quicker canopy closure, reducing the area of bare soil exposed to erosive forces. This enhances soil moisture retention and organic matter accumulation.
  - **Efficient Resource Utilization:** With more plants per unit area, there can be more efficient capture of sunlight and absorption of nutrients and water from the soil, assuming adequate resource availability per plant.
  - **Uniform Competition:** Each plant experiences similar competitive pressure from its neighbors, potentially leading to more uniform growth.
- **Challenges and Sustainability Implications:**
  - **Complexity of Layout:** Laying out a hexagonal pattern is more intricate and time-consuming than a square pattern, requiring precise measurements and skilled labor or specialized GPS equipment.
  - **Mechanization Difficulties:** While some operations (*e.g.*, spraying) can still be performed linearly, inter-row cultivation and harvesting machinery designed for straight, perpendicular passes may be less efficient or require more complex maneuvers in a hexagonal

setup. This can increase operational costs or necessitate specialized machinery.

- **Increased Competition:** The higher plant density, while beneficial for overall yield, can lead to increased competition among individual plants for light, water, and nutrients if resources are limited. This might necessitate higher input levels (fertilizer, irrigation) or result in reduced individual plant growth and yield if not managed properly. This balance is key to its sustainability.

(ii) Modern Planting Patterns

Modern planting patterns move beyond simple geometric arrangements, integrating principles of conservation and ecological engineering to address environmental challenges and enhance system resilience. They often prioritize soil health, water management, and biodiversity.

A. Contour Planting

Contour planting is a transformative planting pattern specifically designed for sloping land. It involves orienting rows of plants along the natural contour lines of the terrain, perpendicular to the slope. This creates a series of miniature dams that intercept runoff and prevent soil erosion.

- ☆ **Benefits:**
  - **Effective Soil Erosion Control:** This is the primary benefit. By slowing down water flow, contour planting dramatically reduces the velocity of runoff, preventing sheet and rill erosion. The plant rows act as barriers, trapping soil particles and promoting infiltration.
  - **Enhanced Water Infiltration and Retention:** The contour lines allow rainwater to soak into the soil rather than running off, increasing soil moisture availability for crops and reducing the need for supplemental irrigation. This is particularly crucial in rain-fed agriculture.
  - **Nutrient Preservation:** By reducing soil erosion, contour planting prevents the loss of valuable topsoil and the nutrients contained within it, maintaining long-term soil fertility.
  - **Microclimate Moderation:** The terraces or ridges created can slightly alter local microclimates, potentially reducing wind speed and soil temperature fluctuations.
- ☆ **Challenges and Sustainability Implications:**
  - **Complex Layout and Implementation:** Laying out contour lines accurately requires surveying skills or precision GPS equipment. The rows are not straight, which can complicate mechanical operations.

- **Variable Row Lengths and Spacing:** The curvature of the land means that row lengths will vary, and maintaining uniform plant spacing can be challenging. This might require more precise operator control for machinery.
- **Risk of Gully Formation:** If contour lines are not properly laid out or if extreme rainfall events occur, concentrated water flow points can still lead to gully erosion, particularly at the ends of rows. Proper design with diversion ditches or grassed waterways is essential.
- **Not Suitable for All Slopes:** While effective for moderate slopes, extremely steep slopes may require more intensive conservation measures like terracing.

B. Strip Cropping

Strip cropping is a highly effective conservation planting pattern that involves cultivating different crops in alternating, relatively narrow strips on the same land. These strips are typically arranged perpendicular to the direction of the prevailing wind or parallel to the contours of a slope.

- **Benefits:**
  - **Superior Erosion Control (Wind and Water):** The alternating strips of erosion-resistant crops (*e.g.*, sod, perennial grasses, small grains) and erosion-susceptible crops (*e.g.*, corn, soybeans, cotton) act as a living barrier. The dense cover crops effectively trap soil particles, slow down wind speed, and reduce water runoff.
  - **Improved Soil Health:** The inclusion of cover crops or perennial strips helps build soil organic matter, improve soil structure, and enhance water infiltration.
  - **Enhanced Biodiversity:** Different crops provide varied habitats and food sources for beneficial insects, pollinators, and wildlife, contributing to overall ecosystem health.
  - **Pest and Disease Management:** The strips can act as physical barriers to pest movement or provide habitat for natural enemies of pests, reducing the reliance on chemical inputs.
  - **Nutrient Cycling:** Leguminous cover crops can contribute nitrogen to the soil, benefiting adjacent cash crops and reducing the need for synthetic fertilizers.
- **Challenges and Sustainability Implications:**
  - **Increased Management Complexity:** Planning and managing multiple crops in strips can be more complex than monoculture, requiring careful consideration of crop compatibility, planting and harvesting times, and machinery requirements.

- **Edge Effects:** The transition zones between strips can create unique microclimates and pest environments that need to be managed.
- **Mechanization Adapting:** While possible, machinery operations may need to be adjusted for the narrower strips, potentially requiring smaller equipment or more passes.
- **Long-Term Planning:** Requires a rotational plan for the different strips to ensure soil health benefits and prevent nutrient imbalances over time.

Planting Designs for Specific Crops (Agroecological Systems)

Beyond general patterns, specific planting designs integrate diverse plant species to leverage ecological interactions, mimicking natural ecosystems to enhance productivity and resilience. These designs are central to agroecological and regenerative agriculture principles.

A. Intercropping

Intercropping involves growing two or more crops simultaneously in the same field. This design maximizes land use efficiency by exploiting different growth habits, resource requirements, and temporal niches of co-planted species.

☆ **Benefits:**

- **Increased Land Use Efficiency (LER > 1):** The Land Equivalent Ratio (LER) often exceeds 1, meaning that a greater total yield can be obtained from intercropped land than from the same area planted with monocultures of the individual crops. Different crops utilize resources (light, water, nutrients) from different soil depths or at different times.
- **Pest and Disease Suppression:** Diversification can confuse pests, provide habitat for natural enemies, or act as trap crops or repellent crops, reducing pest pressure and the need for pesticides.
- **Improved Nutrient Cycling:** Legume-cereal intercropping systems are classic examples, where the legume fixes atmospheric nitrogen, making it available to the non-leguminous companion crop, reducing the need for synthetic nitrogen fertilizers.
- **Reduced Soil Erosion:** Increased ground cover from multiple crops provides better protection against wind and water erosion.
- **Yield Stability and Risk Diversification:** If one crop fails or performs poorly due to environmental stress or pest outbreak, the other crop may still provide a yield, buffering income and food security.
- **Weed Suppression:** A denser, more diverse canopy can outcompete weeds.

☆ **Challenges and Sustainability Implications:**

- **Inter-species Competition:** If not carefully selected and managed, crops can compete for light, water, and nutrients, potentially reducing the yield of one or both species.
- **Increased Management Complexity:** Requires a deeper understanding of crop interactions, staggered planting times, and sometimes separate harvesting methods.
- **Mechanization Difficulties:** Simultaneous harvesting of multiple crops and operating machinery in diverse stands can be challenging, often favoring manual labor or specialized equipment.
- **Allelopathy:** Some plants release biochemicals that can inhibit the growth of neighboring species, which must be considered during selection.

B. Agroforestry

Agroforestry is a dynamic, ecologically based natural resource management system that integrates trees with crops and/or livestock on the same land management unit. It is a highly diverse and adaptable approach, often seen as a cornerstone of sustainable land use.

☆ **Benefits:**

- **Enhanced Productivity and Diversification:** Provides multiple products (food, fodder, fuel, fiber, timber) and services from the same land unit, leading to diversified income streams and increased overall productivity.
- **Carbon Sequestration:** Trees are highly effective at sequestering atmospheric carbon dioxide in their biomass and in the soil, contributing significantly to climate change mitigation.
- **Soil Health Improvement:** Tree roots stabilize soil, reduce erosion, and improve soil structure. Leaf litter and root exudates increase soil organic matter, nutrient cycling, and microbial activity.
- **Biodiversity Conservation:** Creates diverse habitats for a wide range of flora and fauna, enhancing ecosystem biodiversity, including beneficial insects and wildlife.
- **Water Management:** Trees can improve water infiltration, reduce runoff, and recharge groundwater. Riparian buffer strips protect water quality by filtering pollutants.
- **Microclimate Regulation:** Trees provide shade, reduce wind speed, and moderate temperature extremes, creating more favorable conditions for crops and livestock.
- **Resilience to Climate Change:** More diverse systems are often more resilient to environmental shocks like droughts or extreme weather events.

☆ **Challenges and Sustainability Implications:**

- **Long Establishment Period:** Trees take time to mature, meaning initial returns can be delayed, requiring long-term planning and investment.
- **Competition for Resources:** Managing the balance between trees and crops/livestock to minimize competition for light, water, and nutrients is crucial and can be complex.
- **Management Intensity:** Agroforestry systems often require more nuanced and knowledge-intensive management compared to monoculture.
- **Marketing Challenges:** Diversified outputs may require developing multiple market channels for different products.
- **Policy and Land Tenure Issues:** Long-term commitment required for trees can conflict with short-term land tenure systems or unsuitable policy frameworks.

Challenges and Recommendations

While the various planting patterns and designs offer substantial benefits for sustainable plantation management, their adoption is not without hurdles. Successfully integrating these approaches requires addressing practical, economic, and knowledge-based challenges.

Challenges

1. **Complexity and Knowledge Gap:** More sophisticated patterns and designs (*e.g.*, contour planting, intercropping, agroforestry) require a deeper understanding of ecological principles, crop interactions, and site-specific conditions. Farmers and plantation managers may lack the necessary expertise.
2. **Initial Investment:** Laying out complex patterns, acquiring specialized equipment (*e.g.*, GPS-guided planters for contour farming, tools for diverse crop harvesting), or establishing tree components in agroforestry can involve significant upfront capital investment.
3. **Increased Labor Requirements:** Some patterns, particularly intercropping and certain agroforestry systems, may necessitate more manual labor for planting, weeding, and harvesting due to reduced mechanization efficiency or the need for more nuanced management.
4. **Mechanization Challenges:** Standard large-scale farm machinery is often designed for uniform, straight rows. Adapting these machines or acquiring specialized ones for curved rows, heterogeneous strips, or mixed tree-crop systems can be difficult and costly.
5. **Potential for Reduced Individual Crop Yields:** In some intercropping or agroforestry systems, competition between species might lead to a

slightly lower yield for individual crops compared to their monoculture counterparts, even if the overall system productivity (LER) is higher.

6. **Market Access and Diversification:** Producing a wider variety of crops or tree products from diverse systems requires developing multiple market channels, which can be challenging for producers accustomed to specializing in a single commodity.

Recommendations for Sustainable Adoption

To overcome these challenges and fully realize the potential of advanced planting patterns and designs, the following recommendations are crucial:

1. **Leverage Precision Agriculture Technologies:** The advancements in precision agriculture are instrumental in making complex patterns more feasible.
    - **GPS-guided Equipment:** Allows for highly accurate and efficient layout of contour lines, hexagonal patterns, and precise intercropping configurations, significantly reducing labor and errors during planting.
    - **Variable Rate Application:** Enables site-specific application of inputs (fertilizers, water), optimizing resource use and minimizing waste, crucial for the diverse needs within complex systems.
    - **Remote Sensing and Drones:** Provide high-resolution imagery for monitoring plant health, identifying variations, and informing adaptive management decisions across varied planting designs.
2. **Integrate Cover Crops and Strategic Crop Rotations:** These practices are synergistic with virtually all planting patterns.
    - **Improved Soil Health:** Cover crops (planted between or within main crop rows) and well-planned rotations enhance soil organic matter, nutrient cycling, water infiltration, and soil structure, making the land more resilient and productive.
    - **Pest and Disease Management:** Diverse rotations and cover crops break pest and disease cycles, reducing reliance on chemical interventions.
    - **Erosion Control:** Provide continuous ground cover, particularly in patterns like square/rectangular, mitigating their inherent erosion risks.
3. **Adopt Conservation Tillage Practices (No-Till or Reduced-Till Farming):**
    - **Reduced Soil Erosion:** Minimizing soil disturbance significantly reduces wind and water erosion, preserving topsoil and soil structure.

- **Enhanced Soil Health:** Promotes the accumulation of organic matter, improves water holding capacity, and supports vibrant soil microbial communities, which are essential for long-term sustainability.
- **Carbon Sequestration:** No-till practices are highly effective at sequestering carbon in the soil, contributing to climate change mitigation.
- **Fuel and Labor Savings:** Reduces the number of passes required, lowering fuel consumption and labor costs. Conservation tillage complements patterns that maintain ground cover.

4. **Invest in Research, Extension, and Farmer Training:**
   - **Knowledge Transfer:** Develop and disseminate region-specific best practices for implementing and managing various planting designs and patterns.
   - **Capacity Building:** Provide practical training for farmers and plantation managers on techniques, equipment use, and ecological principles.
   - **Adaptive Management:** Encourage continuous learning and adaptation of designs based on local conditions and performance feedback.
5. **Foster Supportive Policies and Economic Incentives:**
   - **Financial Support:** Provide subsidies, grants, or low-interest loans for adopting sustainable planting practices and associated technologies.
   - **Market Development:** Support the development of value chains for diversified products from intercropping and agroforestry systems.
   - **Certification Schemes:** Promote sustainable certification that rewards environmentally responsible plantation management.

Conclusion

Planting patterns and designs are far more than mere arrangements of plants; they are fundamental strategic choices that dictate the ecological footprint, economic viability, and long-term sustainability of plantation systems. The evolution from straightforward traditional patterns, often optimized for simple mechanization but with environmental trade-offs, to modern, ecologically integrated designs marks a critical shift in land management philosophy.

Experiment 10

Natural and Artificial Regeneration

Objective: A Comparative Analysis on Natural and Artificial Regeneration in Agroforestry Systems

Preamble: Agroforestry, the intentional integration of trees and shrubs with agricultural crops and/or livestock, offers a sustainable land-use approach that enhances biodiversity, soil health, and overall ecosystem services. Regeneration, the process of re-establishing trees in an agroforestry system, is crucial for its long-term viability and success. This illustration explores the two primary methods of tree regeneration in agroforestry: natural regeneration and artificial regeneration. We analyze their strengths, weaknesses, applicability in different agroforestry systems, and the factors influencing their effectiveness. This comparative analysis provides insights into informed decision-making for promoting sustainable and productive agroforestry practices.

Keywords: Agroforestry, Natural Regeneration, Artificial Regeneration, Regeneration Techniques, Sustainability, Biodiversity, Ecosystem Services.

Introduction

Agroforestry is gaining increasing recognition as a vital strategy for addressing global challenges such as food security, climate change mitigation, and biodiversity conservation. By strategically combining trees with agricultural land, agroforestry systems harness the synergistic interactions between these components, leading to enhanced productivity, resilience, and environmental benefits. One critical aspect of maintaining and enhancing agroforestry systems is the successful regeneration of trees, ensuring their long-term presence and contributing to the system's overall health and functionality.

Regeneration in agroforestry can occur through two main pathways: natural regeneration, relying on self-seeding and vegetative propagation, and artificial regeneration, involving human intervention through planting or direct seeding. The choice between these methods depends on various factors, including the desired tree species, site conditions, management objectives, and available resources. Understanding the nuances of each approach is crucial for designing and implementing effective agroforestry systems that maximize ecological and economic benefits.

Natural Regeneration in Agroforestry

Natural regeneration refers to the establishment of new trees through natural processes, without direct human intervention in planting. It relies on existing seed sources, either from mature trees within the agroforestry system or from nearby forests. The process involves seed dispersal, germination, seedling establishment, and subsequent growth.

Advantages of Natural Regeneration

- ✰ **Cost-effectiveness:** Natural regeneration generally requires lower initial investment compared to artificial regeneration, as it eliminates the cost of seedlings, planting labor, and related materials.
- ✰ **Adaptability:** Naturally regenerated seedlings are often better adapted to local site conditions, as they have undergone natural selection processes. This can lead to higher survival rates and greater resilience to local pests and diseases.
- ✰ **Genetic Diversity:** Natural regeneration promotes greater genetic diversity within the tree population, enhancing the system's long-term adaptability to changing environmental conditions.
- ✰ **Ecosystem Benefits:** Natural regeneration allows for the natural succession of tree species, promoting biodiversity and contributing to the restoration of natural ecosystems.

Disadvantages of Natural Regeneration

- ✰ **Uncertainty:** The success of natural regeneration can be highly variable and dependent on unpredictable factors such as seed availability, germination rates, and seedling survival.
- ✰ **Slow Process:** Natural regeneration can be a slow process, especially in degraded or disturbed sites. It may take several years for trees to reach a size where they contribute significantly to the agroforestry system.
- ✰ **Limited Species Control:** Natural regeneration may result in the establishment of undesirable tree species, competing with desired species and potentially hindering the overall productivity of the system.

- ☆ **Competition:** Newly established seedlings can face intense competition from existing vegetation, including agricultural crops and weeds, potentially hindering their growth and survival.

Factors Influencing Natural Regeneration

- ☆ **Seed Availability and Dispersal:** The proximity and abundance of mature trees, as well as the effectiveness of seed dispersal mechanisms (wind, water, animals), directly impact the availability of seeds in the agroforestry system.
- ☆ **Seedbed Conditions:** The presence of a suitable seedbed, with adequate moisture, nutrients, and sunlight, is critical for seed germination and seedling establishment. Soil disturbance can often improve seedbed conditions.
- ☆ **Herbivory and Pest Pressure:** Grazing animals and insect pests can significantly reduce seedling survival rates. Protecting seedlings from herbivory is often necessary.
- ☆ **Weed Competition:** Controlling weeds is essential to reduce competition for resources and provide seedlings with adequate access to sunlight, water, and nutrients.
- ☆ **Fire Regimes:** Fire can be devastating to young seedlings, but can also play a role in creating favorable seedbed conditions for certain species. Controlled burns may be used in some systems.

Artificial Regeneration in Agroforestry

Artificial regeneration involves the direct planting or seeding of trees by humans to establish a new generation. This method allows for greater control over species selection, spacing, and timing of tree establishment.

Advantages of Artificial Regeneration

- ☆ **Species Selection:** Artificial regeneration allows for the selection of specific tree species that are well-suited to the agroforestry system's objectives, such as fast-growing species for timber production or nitrogen-fixing species for soil improvement.
- ☆ **Control over Spacing and Arrangement:** Planting allows for precise control over the spacing and arrangement of trees, optimizing light interception, competition, and management practices.
- ☆ **Faster Establishment:** Artificial regeneration can result in faster establishment and growth of trees compared to natural regeneration, as seedlings are typically larger and more robust at the time of planting.
- ☆ **Higher Survival Rates:** With proper planting techniques, site preparation, and post-planting care, artificial regeneration can achieve

higher survival rates compared to natural regeneration, especially in challenging environments.

Disadvantages of Artificial Regeneration

- ✰ **Higher Costs:** Artificial regeneration requires significant financial investment in seedlings, planting labor, site preparation, and ongoing maintenance.
- ✰ **Reduced Adaptability:** Planted seedlings may be less adapted to local site conditions compared to naturally regenerated seedlings, potentially leading to lower survival rates and reduced resilience.
- ✰ **Lower Genetic Diversity:** Planting programs often rely on a limited number of provenances or clones, resulting in reduced genetic diversity within the tree population.
- ✰ **Environmental Impact:** Seedling production and transportation can have environmental impacts, such as greenhouse gas emissions and the depletion of natural resources.

Techniques for Artificial Regeneration

- ✰ **Direct Seeding:** This involves sowing seeds directly into the soil. It is often less expensive than planting seedlings but requires careful site preparation and seed protection from predators.
- ✰ **Bare-Root Planting:** Bare-root seedlings are grown in nurseries and transplanted without soil around their roots. This is a cost-effective method but requires careful handling and planting to ensure survival.
- ✰ **Containerized Seedlings:** Containerized seedlings are grown in pots or containers, allowing for a more controlled environment and reduced transplant shock. This method is more expensive but often results in higher survival rates.
- ✰ **Grafting and Cloning:** These techniques can be used to propagate superior trees with desirable traits, such as disease resistance or high yield. They are particularly useful for establishing orchards and other high-value agroforestry systems.

Comparative Analysis and Considerations for Agroforestry Systems

The choice between natural and artificial regeneration in agroforestry depends on a variety of factors, including the specific agroforestry system, site conditions, management objectives, and available resources. Table 1 provides a summary comparison of the two methods.

Table 1: Comparative Analysis of Natural and Artificial Regeneration in Agroforestry

Feature	*Natural Regeneration*	*Artificial Regeneration*
Cost	Lower	Higher
Adaptability	Greater	Lower
Genetic Diversity	Higher	Lower
Establishment Speed	Slower	Faster
Species Control	Lower	Higher
Survival Rates	Potentially Lower	Potentially Higher
Labor Requirements	Lower	Higher
Management Intensity	Lower	Higher

Considerations for Specific Agroforestry Systems

- **Silvopastoral Systems:** Natural regeneration may be suitable in established silvopastoral systems where existing trees are present and grazing pressure is managed. However, artificial regeneration may be necessary in degraded areas or when introducing new tree species.
- **Alley Cropping:** Artificial regeneration is often preferred in alley cropping systems to ensure proper spacing and arrangement of trees within the alleys.
- **Forest Farming:** Natural regeneration can be a valuable tool in forest farming systems, allowing for the natural succession of tree species and promoting biodiversity. However, artificial regeneration may be used to introduce specific understory crops or enhance the growth of desired tree species.
- **Home Gardens:** Both natural and artificial regeneration can be used in home gardens, depending on the specific needs and preferences of the gardener.

Conclusion

Both natural and artificial regeneration play important roles in maintaining and enhancing agroforestry systems. Natural regeneration offers a cost-effective and ecologically sound approach, promoting biodiversity and adapting to local site conditions. Artificial regeneration provides greater control over species selection, spacing, and establishment, leading to faster growth and higher survival rates.

The optimal choice between natural and artificial regeneration depends on a careful assessment of the specific agroforestry system, site conditions, management objectives, and available resources. A well-designed agroforestry system may even incorporate both strategies, utilizing natural regeneration to promote biodiversity and resilience, while employing artificial regeneration to introduce specific species or enhance productivity. By understanding the strengths and weaknesses of each approach, agroforestry practitioners can make informed decisions that promote sustainable and productive land-use practices, contributing to both ecological and economic benefits.

Further Research

Future research should focus on:

- ☆ Developing improved techniques for managing natural regeneration in agroforestry systems.
- ☆ Evaluating the long-term economic and ecological impacts of different regeneration strategies.
- ☆ Investigating the role of genetic diversity in the resilience of agroforestry systems.
- ☆ Developing cost-effective and environmentally friendly methods for artificial regeneration.
- ☆ Assessing the impact of climate change on the success of natural and artificial regeneration in different agroforestry systems.

By continuing to research and improve our understanding of tree regeneration in agroforestry, we can ensure the long-term sustainability and productivity of these valuable land-use systems.

Experiment 11

Design and Diagnostic Survey of Agroforestry System

Objective: An Integrated Approach to Sustainable Agriculture and Forestry

Preamble: Agroforestry systems offer a sustainable and integrated approach to agriculture and forestry, combining trees, shrubs, crops, and livestock to create diverse, productive, and resilient landscapes. The design of agroforestry systems involves a thorough assessment of site-specific factors, goal setting, component selection, spatial arrangement, and management practices. A diagnostic survey, on the other hand, is a systematic assessment of an existing agroforestry system to evaluate its performance, identify constraints, and recommend improvements. By integrating ecological, social, and economic factors, these approaches can help create agroforestry systems that contribute to food security, climate change mitigation, and rural development.

To promote the adoption and scaling-up of agroforestry practices worldwide, it is crucial to invest in research, training, and extension services. This will enable farmers, foresters, and other stakeholders to better understand the benefits of agroforestry and develop the skills and knowledge necessary to design, implement, and manage these systems effectively. Additionally, policies and incentives should be put in place to encourage the adoption of agroforestry practices, particularly in areas where land degradation, food insecurity, and climate change pose significant challenges.

In summary, the design and diagnostic survey of agroforestry systems are essential components of sustainable land-use management. By integrating ecological, social, and economic factors, these approaches can help create diverse, productive, and resilient landscapes that contribute to the well-being of both people and the environment. As such, it is crucial to continue promoting and supporting agroforestry practices as an integral part of sustainable agriculture and forestry.

Introduction

Agroforestry is a sustainable land-use practice that integrates trees and shrubs with crops and/or livestock to create diverse, productive, and resilient landscapes. This illustration aims to provide an overview of the design and diagnostic survey of agroforestry systems, highlighting the importance of integrating ecological, social, and economic factors in the development and management of these systems.

Design of Agroforestry Systems

The design of agroforestry systems involves the selection and arrangement of tree, shrub, crop, and livestock components to optimize the benefits derived from their interactions. The following are key elements to consider in the design process:

1. Site assessment: Analyze site-specific factors such as soil type, topography, climate, and existing vegetation to determine the most suitable agroforestry practices for the location.
2. Goal setting: Establish clear objectives for the agroforestry system, considering factors such as food security, income generation, environmental conservation, and social well-being.
3. Component selection: Choose appropriate tree, shrub, crop, and livestock species based on their compatibility, ecological functions, and potential benefits.
4. Spatial arrangement: Determine the optimal spatial arrangement of components to maximize resource use efficiency, minimize competition, and enhance ecological interactions.
5. Management practices: Develop a comprehensive management plan that includes guidelines for planting, pruning, fertilization, pest and disease control, and harvesting.

Diagnostic Survey of Agroforestry Systems

A diagnostic survey is a systematic assessment of an existing agroforestry system to evaluate its performance, identify constraints, and recommend improvements. The following are key components of a diagnostic survey:

1. Data collection: Gather information on the site's physical, biological, and socio-economic characteristics, as well as the performance of the agroforestry system. This may involve interviews with stakeholders, field observations, and the collection of relevant data.

2. Analysis and interpretation: Analyze the collected data to identify patterns, trends, and relationships that can help explain the system's performance. This may involve the use of statistical techniques, geographic information systems (GIS), and remote sensing.
3. Identification of constraints and opportunities: Based on the analysis, identify the main constraints and opportunities that affect the performance of the agroforestry system. These may include biophysical factors (*e.g.*, soil fertility, pests, and diseases), socio-economic factors (*e.g.*, land tenure, market access, and labor availability), and institutional factors (*e.g.*, policies, laws, and governance).
4. Recommendations for improvement: Develop a set of recommendations to address the identified constraints and opportunities, considering both short-term and long-term goals. These may include changes in management practices, the introduction of new species, or the development of new markets and value chains.

Conclusion

The design and diagnostic survey of agroforestry systems are essential tools for the sustainable management of agricultural and forestry landscapes. By integrating ecological, social, and economic factors, these approaches can help create diverse, productive, and resilient agroforestry systems that contribute to food security, climate change mitigation, and rural development. As such, it is crucial to invest in research, training, and extension services to promote the adoption and scaling-up of agroforestry practices worldwide.

Experiment 12

Evaluation of Agroforestry System in different Agroclimatic Zones

Objective: Evaluation of Agroforestry Systems in Different Agroclimatic Zones

Preamble: Agroforestry, the intentional integration of trees and shrubs with crops and/or livestock, holds immense potential for enhancing agricultural productivity, improving environmental sustainability, and building resilience to climate change. Its effectiveness, however, is highly dependent on the specific agroclimatic zone, which dictates the suitability of different agroforestry practices and the potential benefits they can deliver. This illustration examines the evaluation of agroforestry systems across diverse agroclimatic zones, focusing on the methodologies used, the key performance indicators assessed, and the challenges encountered in quantifying the impacts of agroforestry. The review highlights the need for tailored agroforestry designs and context-specific evaluation frameworks to maximize the benefits and overcome the limitations of these systems in different environmental settings.

Keywords: Agroforestry, Agroclimatic Zones, Evaluation, Performance Indicators, Sustainability, Productivity, Resilience, Climate Change, Environmental Impact.

Introduction

Agroforestry has emerged as a promising land management strategy that blends the ecological benefits of forests with the productivity of agriculture. It offers a multitude of advantages, including soil conservation, carbon sequestration, biodiversity enhancement, improved water management, and diversification of income sources for farmers. The success of agroforestry, however, hinges on its adaptability to the unique biophysical and socio-economic conditions of specific agroclimatic zones.

Agroclimatic zones are defined regions characterized by similar climatic conditions and agricultural potential. These zones influence the suitability of different tree species, crop types, and livestock breeds, as well as farming practices. Therefore, the evaluation of agroforestry systems must consider the specific context of each agroclimatic zone to accurately assess its performance and identify opportunities for improvement.

This illustration aims to provide a comprehensive overview of the evaluation of agroforestry systems across diverse agroclimatic zones. It explores the methodologies used to assess agroforestry practices, the key performance indicators employed, and the challenges faced in quantifying the impacts of these systems. The illustration also highlights the importance of developing context-specific agroforestry designs and evaluation frameworks to optimize the benefits of agroforestry in different environments.

Agroclimatic Zones and Their Impact on Agroforestry Systems

Agroclimatic zones are typically classified based on factors such as temperature, rainfall, humidity, solar radiation, and growing season length. Some of the major agroclimatic zones include:

- ✰ **Tropical Rainforests:** Characterized by high rainfall and humidity, these zones are suitable for diverse agroforestry systems involving shade-tolerant crops like cocoa, coffee, and spices under a canopy of trees.
- ✰ **Tropical Savannas:** Marked by distinct wet and dry seasons, these zones favor agroforestry systems that integrate drought-resistant trees and shrubs with crops like millet, sorghum, and groundnuts.
- ✰ **Temperate Regions:** Experiencing distinct seasons with moderate rainfall, these zones allow for the integration of fruit trees, timber trees, and livestock with crops like wheat, barley, and maize.
- ✰ **Mediterranean Climates:** Characterized by hot, dry summers and mild, wet winters, agroforestry systems in these zones often involve olive trees, vineyards, and grazing livestock.
- ✰ **Arid and Semi-arid Regions:** Experiencing low rainfall and high evapotranspiration, these zones require agroforestry systems that prioritize water conservation and the use of drought-tolerant species like acacia and prosopis.
- ✰ **Boreal Forests:** Characterized by cold winters and short growing seasons, agroforestry in these zones often involves integrating forest management practices with the cultivation of berries and medicinal plants.

The choice of agroforestry practices, the species selected, and the management strategies employed must be tailored to the specific agroclimatic conditions of each zone. For instance, alley cropping might be highly effective in a humid tropical region, but less suitable in an arid environment due to competition for water. Similarly, the selection of nitrogen-fixing trees is crucial in nutrient-poor soils, while the use of windbreaks is particularly important in windy regions.

Evaluation Methodologies for Agroforestry Systems

Several methodologies are employed to evaluate the performance of agroforestry systems across different agroclimatic zones. These methods can be broadly categorized into biophysical and socio-economic assessments.

Biophysical Assessments

- **Productivity Measurement:** This involves quantifying the yields of crops, timber, and livestock produced under agroforestry systems. Data are collected through plot-level measurements, farmer surveys, and statistical analysis.
- **Soil Properties Analysis:** This assesses the impact of agroforestry on soil fertility, organic matter content, soil structure, and water infiltration. Soil samples are collected and analyzed in the laboratory.
- **Carbon Sequestration Measurement:** This quantifies the amount of carbon stored in trees, soil, and other biomass components of the agroforestry system. Methods include tree inventory, biomass estimation, and soil carbon analysis.
- **Water Use Efficiency Assessment:** This evaluates the impact of agroforestry on water availability and use by crops and trees. Techniques include soil moisture monitoring, evapotranspiration measurement, and hydrological modeling.
- **Biodiversity Assessment:** This quantifies the diversity and abundance of plant and animal species within the agroforestry system. Methods include species inventories, bird counts, and insect trapping.

Socio-economic Assessments

- **Cost-Benefit Analysis:** This compares the economic costs and benefits of agroforestry systems, including labor inputs, input costs, and revenue generated from different products.
- **Farmer Surveys:** This gathers information from farmers about their perceptions, attitudes, and practices related to agroforestry. Surveys can assess factors such as adoption rates, management practices, and satisfaction levels.

- **Participatory Rural Appraisal (PRA):** This involves engaging local communities in the evaluation process, using participatory techniques to gather information about their knowledge, needs, and priorities related to agroforestry.
- **Livelihood Analysis:** This assesses the impact of agroforestry on household income, food security, and overall well-being.
- **Market Analysis:** This examines the market opportunities for agroforestry products, including timber, fruits, and other non-timber forest products.

Combining biophysical and socio-economic assessments is crucial for a comprehensive evaluation of agroforestry systems. This integrated approach provides a holistic understanding of the environmental and socio-economic impacts of agroforestry, informing decision-making and promoting the sustainable adoption of these practices.

Key Performance Indicators for Agroforestry Systems

The selection of appropriate performance indicators is essential for evaluating the success of agroforestry systems. The specific indicators used will vary depending on the goals of the evaluation and the characteristics of the agroclimatic zone. Some commonly used indicators include:

- **Crop Yield:** Quantifies the productivity of crops grown in association with trees.
- **Tree Growth Rate:** Measures the increase in tree height, diameter, and biomass over time.
- **Soil Organic Matter Content:** Indicates the level of organic matter in the soil, which is a key indicator of soil health and fertility.
- **Soil Erosion Rate:** Measures the amount of soil lost due to erosion, reflecting the effectiveness of agroforestry in soil conservation.
- **Water Infiltration Rate:** Indicates the rate at which water penetrates the soil, reflecting the impact of agroforestry on water availability.
- **Carbon Sequestration Rate:** Measures the amount of carbon stored in trees, soil, and other biomass components of the agroforestry system.
- **Biodiversity Index:** Quantifies the diversity and abundance of plant and animal species within the agroforestry system.
- **Farmer Income:** Measures the income generated by farmers from agroforestry products.
- **Adoption Rate:** Indicates the percentage of farmers who are adopting agroforestry practices.
- **Farmer Satisfaction:** Measures the level of satisfaction among farmers with agroforestry systems.

- ☆ **Return on Investment (ROI):** Measures the profitability of the agroforestry investment.
- ☆ **Household Food Security:** Assesses the availability and accessibility of food for the household.

Challenges in Evaluating Agroforestry Systems

Evaluating agroforestry systems presents several challenges, including:

- ☆ **Complexity of Interactions:** Agroforestry involves complex interactions between trees, crops, livestock, and the environment. It can be difficult to isolate the specific effects of agroforestry on different components of the system.
- ☆ **Long Time Horizons:** Trees can take several years to mature, making it difficult to assess the long-term impacts of agroforestry within a reasonable timeframe.
- ☆ **Spatial Variability:** Agroclimatic conditions, soil properties, and management practices can vary significantly across small areas, making it challenging to generalize findings across different locations.
- ☆ **Data Availability:** Reliable data on crop yields, tree growth, soil properties, and socio-economic indicators may be limited, particularly in developing countries.
- ☆ **Attribution Issues:** It can be difficult to attribute changes in productivity or environmental conditions solely to agroforestry, as other factors such as climate change, land degradation, and policy interventions may also play a role.
- ☆ **Standardization of Methods:** Lack of standardized evaluation methods and indicators makes it challenging to compare results across different studies and agroclimatic zones.
- ☆ **Scaling Up:** Evaluating demonstration plots or small-scale agroforestry systems may not accurately reflect the impacts of large-scale implementation.

Case Studies: Examples of Agroforestry Evaluation in Different Agroclimatic Zones

To illustrate the application of evaluation methodologies in different agroclimatic zones, consider the following examples:

- ☆ **Humid Tropics (Indonesia):** Studies on cocoa-based agroforestry systems in Indonesia have used productivity measurements, soil analysis, and carbon sequestration measurements to evaluate the impact of these systems on cocoa yields, soil fertility, and carbon storage. Researchers have found that cocoa agroforestry systems can significantly improve soil health and carbon sequestration compared to monoculture cocoa plantations.

- ☆ **Semi-arid Regions (Sahel):** Evaluations of parkland agroforestry systems in the Sahel region of Africa have focused on the impact of trees on crop yields, soil moisture, and livestock productivity. Farmer surveys and participatory rural appraisal techniques have been used to assess farmer perceptions of these systems. Studies have shown that parkland agroforestry can enhance crop yields, improve soil fertility, and provide valuable fodder for livestock.
- ☆ **Temperate Regions (Europe):** Research on silvoarable systems in Europe has focused on the impact of trees on crop yields, biodiversity, and carbon sequestration. Researchers have used productivity measurements, biodiversity assessments, and carbon sequestration measurements to evaluate the performance of these systems. Studies have shown that silvoarable systems can enhance biodiversity, improve soil health, and sequester carbon.
- ☆ **Mediterranean Climate (Spain):** Studies of Dehesa and Montado systems (oak woodlands integrating livestock grazing) have explored the impact on biodiversity, soil health, and carbon sequestration. Assessments combining ecological analysis and economic modelling have been employed to understand the tradeoffs and synergies in these systems.

Improving Agroforestry Evaluation: Recommendations

To improve the evaluation of agroforestry systems, the following recommendations are offered:

- ☆ **Develop Standardized Evaluation Methods:** Develop standardized methods and indicators for evaluating agroforestry systems, allowing for more consistent and comparable results across different studies and agroclimatic zones.
- ☆ **Strengthen Data Collection Efforts:** Invest in strengthening data collection efforts, ensuring that reliable data on crop yields, tree growth, soil properties, and socio-economic indicators are readily available.
- ☆ **Use Integrated Assessment Approaches:** Employ integrated assessment approaches that combine biophysical and socio-economic assessments, providing a holistic understanding of the impacts of agroforestry.
- ☆ **Incorporate Long-Term Monitoring:** Establish long-term monitoring programs to track the impacts of agroforestry over time, allowing for a more accurate assessment of its sustainability.

- **Promote Participatory Evaluation:** Promote participatory evaluation approaches that involve local communities in the evaluation process, ensuring that their knowledge, needs, and priorities are taken into account.
- **Use Modeling and Simulation:** Utilize modeling and simulation techniques to predict the long-term impacts of agroforestry under different climate change scenarios.
- **Address Attribution Issues:** Develop methods for addressing attribution issues, allowing for a more accurate determination of the specific impacts of agroforestry.
- **Focus on Scaling Up:** Conduct research on scaling up agroforestry, identifying the factors that influence the successful adoption and implementation of these practices at larger scales.
- **Context-Specific Design and Evaluation:** Emphasize the importance of tailoring agroforestry designs and evaluation frameworks to the specific agroclimatic zone and socio-economic context.

Conclusion

Agroforestry offers a promising pathway towards sustainable agriculture and environmental conservation. However, its effectiveness is highly dependent on the agroclimatic zone and the specific design and management practices employed. The evaluation of agroforestry systems requires a comprehensive understanding of the biophysical and socio-economic interactions between trees, crops, and livestock. By using appropriate evaluation methodologies, selecting relevant performance indicators, and addressing the challenges associated with these assessments, we can gain valuable insights into the potential of agroforestry to enhance agricultural productivity, improve environmental sustainability, and build resilience to climate change in diverse agroclimatic zones. Tailored approaches are critical for maximizing benefits and ensuring the long-term success of agroforestry. Further research and development efforts are needed to refine evaluation methods, promote the adoption of best practices, and scale up the implementation of agroforestry systems around the world.

Experiment 13

Exposure Visit to Prevailing Agroforestry Systems of the State and Related Important Institutions

Objective: Exposure Visit to Prevailing Agroforestry Systems and Related Institutions in the State

Introduction

Agroforestry is a land use management approach that combines agriculture and forestry to create sustainable and productive landscapes. It is a vital component of agriculture and natural resource management in many parts of the world, including the state. This illustration presents an exposure visit to prevailing agroforestry systems in the state and related important institutions.

Agroforestry Systems in the State

The state has a diverse range of agroforestry systems, including silvopastoral, agrosilvicultural, and agroforest-fishery systems. These systems are practiced by smallholder farmers, who integrate trees, crops, and livestock on the same piece of land to optimize land use, increase productivity, and promote sustainability.

One of the most prevalent agroforestry systems in the state is the silvopastoral system, where trees are integrated with livestock farming. The trees provide shade, fodder, and wood for fuel and construction, while the livestock provide manure for the trees and crops. The system is widely practiced in the state's highland areas, where the terrain is not suitable for intensive agriculture.

Another common agroforestry system in the state is the agrosilvicultural system, where trees are integrated with crops. The trees provide shade, soil conservation, and wood for fuel and construction, while the crops provide food and income for the farmers. The system is widely practiced in the state's lowland areas, where the soil is fertile and suitable for intensive agriculture.

The agroforest-fishery system is a unique agroforestry system in the state, where trees are integrated with fish farming. The trees provide shade, wood for fuel and construction, and habitat for fish, while the fish provide protein and income for the farmers. The system is widely practiced in the state's coastal areas, where the soil is saline and unsuitable for agriculture.

Related Institutions

Several institutions in the state are dedicated to promoting and supporting agroforestry practices. The State Forest Department is responsible for promoting agroforestry and providing technical support to farmers. The department has established demonstration plots and training centers to showcase the benefits of agroforestry and train farmers on best practices.

The State Agriculture Department is another important institution that supports agroforestry practices. The department provides extension services, training, and financial support to farmers to promote agroforestry. The department has also established demonstration plots and training centers to promote agroforestry practices.

Non-governmental organizations (NGOs) and community-based organizations (CBOs) also play a crucial role in promoting agroforestry practices in the state. These organizations provide training, extension services, and financial support to farmers to promote agroforestry. They also advocate for policies and programs that support agroforestry practices.

Conclusion

The exposure visit to prevailing agroforestry systems and related institutions in the state has provided valuable insights into the importance and potential of agroforestry practices. The state has a diverse range of agroforestry systems that are widely practiced by smallholder farmers. Several institutions, including the State Forest Department, State Agriculture Department, NGOs, and CBOs, are dedicated to promoting and supporting agroforestry practices. The visit has highlighted the need for continued support and investment in agroforestry practices to promote sustainability, productivity, and resilience in the state's agriculture and natural resource management.

This comprehensive exposure visit to prevailing agroforestry systems and key institutions – whether at the state, national, or international level – has provided profoundly valuable insights into the multifaceted importance and immense potential of agroforestry practices having following features:

1. **Practical Application and Diversity:** Witnessing firsthand the diverse range of agroforestry models in operation (*e.g.*, alley cropping, silvopastoral systems, homegardens, windbreaks) across different

ecological zones and farm sizes. This includes understanding the specific tree and crop combinations that are most successful and why.

2. **Economic Viability:** Gaining a clearer appreciation for the diversified income streams agroforestry provides to farmers, through timber, non-timber forest products, fodder, fruits, and increased crop yields due to improved soil health and microclimates. Understanding the market linkages and value chain aspects.
3. **Ecological Benefits:** Observing the tangible environmental advantages, such as enhanced biodiversity, soil conservation, improved water retention, carbon sequestration, reduced erosion, and the creation of wildlife habitats.
4. **Social Impact and Livelihoods:** Understanding how agroforestry contributes to food security, rural employment, community resilience, and improved living standards for farming communities, particularly in marginalized areas.
5. **Challenges and Solutions:** Learning about the practical challenges faced by farmers and practitioners (*e.g.*, long gestation periods for trees, market access, policy barriers, technical knowledge gaps) and innovative solutions being implemented.
6. **Institutional Support and Policy Frameworks:** Understanding the critical role of the visited institutions – be they research organizations (*e.g.*, ICRAF, local agricultural universities), government departments, NGOs, or farmer cooperatives – in research, development, extension, policy advocacy, and capacity building. This provides a holistic view of the ecosystem supporting agroforestry.
7. **Best Practices and Innovations:** Identifying successful management techniques, innovative technologies (*e.g.*, improved germplasm, water-saving irrigation in agroforestry settings), and sustainable approaches that can be replicated or adapted elsewhere.
8. **Scaling Potential:** Gaining a perspective on how different agroforestry models can be scaled up to benefit larger populations and contribute to national and international sustainable development goals.

In essence, the visit transformed theoretical knowledge into practical understanding, highlighting agroforestry's crucial role in achieving sustainable agriculture, climate change mitigation, and rural development.

In the state of California, the University of California (UC) has several institutions that have been actively involved in agroforestry research and extension. These include the UC Davis Department of Plant Sciences, the UC Agriculture and Natural Resources (ANR), and the UC Cooperative Extension (UCCE) county offices.

Nationally, the USDA National Agroforestry Center (NAC) is a key institution promoting agroforestry practices and providing resources and support to farmers and landowners. Additionally, the American Forest Foundation (AFF) and the National Association of Conservation Districts (NACD) are organizations that advocate for sustainable land management practices, including agroforestry.

Internationally, the World Agroforestry Centre (ICRAF) is a leading research institution focused on agroforestry, with a global network of partners and collaborators. The Food and Agriculture Organization of the United Nations (FAO) also supports agroforestry as a sustainable land management practice and provides resources and guidance to countries worldwide. Other notable international organizations include the International Union for Conservation of Nature (IUCN) and the Consultative Group on International Agricultural Research (CGIAR).

These institutions have provided valuable insights into the importance and potential of agroforestry practices, highlighting their role in enhancing agricultural productivity, improving soil health, increasing climate resilience, and providing ecosystem services such as carbon sequestration, water conservation, and habitat for wildlife. Their research, outreach, and policy advocacy have contributed to the growing adoption of agroforestry practices by farmers and landowners worldwide.

Agroforestry practices are gaining significant recognition for their potential in sustainable land management, climate change mitigation, and enhancing rural livelihoods. Numerous institutions at national, state, and international levels have contributed valuable insights and efforts in this domain.

Here's a list of related institutions:

I. Institutions in India (National Level)

1. Indian Council of Forestry Research and Education (ICFRE)

- ✰ **Role:** An apex body under MoEFCC, primarily responsible for forestry research and education in India. Many of its institutes conduct extensive research on agroforestry systems suitable for different agro-climatic zones.
- ✰ **Key Institutes under ICFRE relevant to Agroforestry:**
  - ❒ **Forest Research Institute (FRI), Dehradun:** A premier institution for forestry research and education, with significant work in agroforestry.
  - ❒ **Institute of Forest Genetics and Tree Breeding (IFGTB), Coimbatore:** Focuses on tree improvement, including species suitable for agroforestry.

- **Institute of Wood Science and Technology (IWST), Bengaluru:** Researches on wood products from agroforestry trees.
- **Arid Forest Research Institute (AFRI), Jodhpur:** Focuses on agroforestry systems for arid and semi-arid regions.
- **Rain Forest Research Institute (RFRI), Jorhat:** Works on agroforestry in humid tropics.

2. Indian Council of Agricultural Research (ICAR)

- **Role:** The apex body for agricultural research and education in India. It supports research on integrated farming systems, including agroforestry.
- **Key Institute under ICAR:**
  - **Central Agroforestry Research Institute (CAFRI), Jhansi:** This is a dedicated ICAR institute specifically for agroforestry research, development, and policy advocacy in India. It's a crucial institution providing insights into various agroforestry models.

3. Ministry of Environment, Forest and Climate Change (MoEFCC)

- **Role:** Policy formulation, regulation, and protection of forests, wildlife, and environment in India. It oversees national forest policies, including those promoting agroforestry.
- **National Agroforestry Policy, 2014:** Launched by the MoEFCC, it's a landmark policy aimed at promoting agroforestry for tree farming, poverty alleviation, and environmental benefits.

4. Ministry of Agriculture and Farmers Welfare, Government of India

- **Role:** Develops and implements policies and programs for agricultural development, promoting practices like integrated farming systems and diversification, which include agroforestry components.

5. National Medicinal Plants Board (NMPB)

- **Role:** Promotes cultivation of medicinal plants, often integrated into agroforestry systems, contributing to both income generation and biodiversity conservation.

6. Agricultural Universities (various across India)

- Many State Agricultural Universities (SAUs) and Deemed Universities conduct research, offer courses, and provide extension services related to agroforestry. Examples include G.B. Pant University of Agriculture and Technology, Punjab Agricultural University, Tamil Nadu Agricultural University, *etc.*

II. Institutions in West Bengal (State Level)

1. West Bengal Forest Department

- ☆ **Role:** Responsible for the management and conservation of forests in the state. It implements various afforestation, social forestry, and agroforestry schemes, including promotion of tree planting on private lands. It has provided insights into suitable species and models for the state's diverse agro-climatic zones.

2. Bidhan Chandra Krishi Viswavidyalaya (BCKV), Kalyani

- ☆ **Role:** A prominent agricultural university in West Bengal. It has departments and research programs focused on forestry, agroforestry, and horticulture, conducting studies on suitable tree-crop combinations, soil health, and socio-economic aspects of agroforestry in the state.

3. Uttar Banga Krishi Viswavidyalaya (UBKV), Cooch Behar

- ☆ **Role:** Another key agricultural university in West Bengal, particularly focusing on the specific agricultural and forestry needs of the North Bengal region. It conducts research and extension activities relevant to agroforestry in the sub-Himalayan and Terai zones.

4. Department of Agriculture, Government of West Bengal

- ☆ **Role:** Implements agricultural development programs and schemes in the state, often promoting diversification and sustainable farming practices that include agroforestry.

5. West Bengal Biodiversity Board

- ☆ **Role:** While not exclusively focused on agroforestry, it's relevant for insights into maintaining biodiversity within agroforestry landscapes and promoting native species.

6. Local NGOs and Community-Based Organizations

- ☆ Numerous smaller Non-Governmental Organizations (NGOs) and Self-Help Groups (SHGs) across West Bengal work closely with farmers on the ground, promoting and implementing various agroforestry models tailored to local conditions. They often bridge the gap between research institutions and farming communities.

III. International Institutions

1. World Agroforestry Centre (ICRAF)

- ☆ **Role:** Headquartered in Nairobi, Kenya, ICRAF is the leading global research organization dedicated to agroforestry. It conducts cutting-edge research, provides policy advice, and disseminates knowledge on the science and practice of agroforestry worldwide. Its insights are foundational to understanding the potential of agroforestry. ICRAF is part of the CGIAR global research partnership.

2. Food and Agriculture Organization (FAO) of the United Nations

- **Role:** FAO provides global leadership in efforts to defeat hunger and improve nutrition and food security. It promotes sustainable land use practices, including agroforestry, through policy support, technical assistance, and capacity building in member countries.

3. Centre for International Forestry Research (CIFOR)

- **Role:** A non-profit scientific institution that conducts research on the most pressing challenges facing forests and people in tropical landscapes. While focusing broadly on forestry, its work often overlaps with and informs agroforestry practices, especially concerning policy, governance, and climate change.

4. International Fund for Agricultural Development (IFAD)

- **Role:** A specialized agency of the United Nations that invests in rural people, empowering them to increase their food security, improve the nutrition of their families and increase their incomes. IFAD often funds projects that incorporate agroforestry as a key component for rural development and climate resilience.

5. Other CGIAR Centers

- While ICRAF is the primary agroforestry center, other CGIAR centers (*e.g.*, International Rice Research Institute - IRRI, International Maize and Wheat Improvement Center - CIMMYT) may also explore intercropping or tree integration within their specific crop mandates, contributing indirectly to agroforestry knowledge.

These institutions collectively contribute through research, policy development, education, extension, and on-the-ground implementation, providing a wealth of insights into the importance, potential, and practical application of agroforestry practices for diverse ecological and socio-economic contexts.

Experiment 14

Virtual Visit of Agroforestry Models Developed by ICAR-IGFRI, ICFRE

Preamble: Agroforestry, the integrated management of trees and shrubs with agricultural crops and/or livestock, offers a promising pathway towards sustainable land management. The Indian Council of Agricultural Research - Indian Grassland and Fodder Research Institute (ICAR-IGFRI) and the Indian Council of Forestry Research and Education (ICFRE) are leading research institutions in India dedicated to promoting agroforestry. This illustration explores the potential benefits of virtual visits to the agroforestry models developed by these institutions. Virtual tours offer accessible and cost-effective methods for disseminating valuable knowledge, promoting best practices, and fostering wider adoption of agroforestry techniques among farmers, researchers, policymakers, and students. The illustration examines the key components of effective virtual visits, the advantages they offer, and the challenges involved in their implementation, highlighting their potential to revolutionize agroforestry education and extension.

Keywords: Agroforestry, ICAR-IGFRI, ICFRE, Virtual Visits, Sustainable Land Management, Knowledge Dissemination, Extension, Education.

Introduction

Agroforestry systems are increasingly recognized for their multifarious benefits, ranging from enhanced soil fertility and biodiversity to improved livelihoods and climate change mitigation. As pressures on land resources intensify due to population growth and climate change, the adoption of sustainable agricultural practices like agroforestry becomes crucial. In India, institutions like ICAR-IGFRI and ICFRE play a pivotal role in developing and demonstrating diverse agroforestry models suitable for varying agro-climatic zones and socio-economic contexts.

However, effectively disseminating the knowledge generated by these institutions to a wider audience remains a significant challenge. Traditional methods of extension, such as field visits and on-site training, are often resource-intensive, time-consuming, and limited in their reach. Virtual visits, leveraging advancements in technology, offer a promising alternative for showcasing agroforestry models and promoting their adoption. By offering interactive and immersive experiences, virtual visits can bridge the gap between research and practice, enabling stakeholders to learn about successful agroforestry implementations regardless of their geographical location.

ICAR-IGFRI and ICFRE as Leaders in Agroforestry Research and Development

ICAR-IGFRI

The Indian Grassland and Fodder Research Institute (ICAR-IGFRI), Jhansi, is a premier institution focused on improving the productivity and sustainability of grasslands and fodder crops. ICAR-IGFRI has developed several agroforestry models that integrate fodder trees, grasses, and arable crops for enhanced livestock production and soil conservation. These models often focus on silvi-pastoral systems, integrating trees with livestock grazing, and agri-silvi-pastoral systems, combining crops, trees, and livestock.

ICFRE

The Indian Council of Forestry Research and Education (ICFRE), Dehradun, is an apex body for forestry research and education in India. Under its umbrella, several research institutes focus on various aspects of forestry, including agroforestry. ICFRE and its constituent institutes have developed diverse agroforestry models incorporating timber trees, fruit trees, medicinal plants, and other valuable species with agricultural crops. Their research emphasizes the ecological and economic benefits of agroforestry, promoting its adoption for land restoration, biodiversity conservation, and livelihood enhancement.

The Concept of Virtual Visits to Agroforestry Models

Virtual visits to agroforestry models involve utilizing digital technologies like virtual reality (VR), augmented reality (AR), 360° photography, and interactive online platforms to provide remote access to real-world agroforestry systems. These virtual tours can be designed to mimic the experience of physically visiting the model, allowing users to:

- ✰ **Explore the layout and design of the agroforestry system:** Users can virtually walk through the model, examine the spatial arrangement of trees, crops, and livestock, and understand the overall design principles.

- **Learn about the species composition and management practices:** Interactive elements can provide information on the different tree species, crop varieties, and livestock breeds used in the agroforestry system, as well as details on planting techniques, pruning methods, irrigation strategies, and pest management practices.
- **Access data and information on the performance of the system:** Virtual visits can incorporate data visualizations showcasing the yields of crops and livestock, the growth rates of trees, the soil health indicators, and the economic returns of the agroforestry system.
- **Interact with experts and ask questions:** Live video conferencing or online forums can facilitate interaction between users and experts in agroforestry, allowing them to ask questions, seek clarification, and receive personalized advice.
- **View multimedia content such as videos, photos, and animations:** Visual aids can enhance the learning experience and provide a deeper understanding of the agroforestry system.

Advantages of Virtual Visits

Virtual visits offer numerous advantages over traditional field visits:

- **Accessibility:** Virtual visits can reach a wider audience, including individuals who may be unable to travel to the site due to physical limitations, geographical constraints, or cost considerations.
- **Cost-effectiveness:** Virtual visits eliminate travel costs, accommodation expenses, and other associated expenses, making them a more affordable option for many stakeholders.
- **Time-efficiency:** Virtual visits can be conducted at any time and from any location, allowing users to learn at their own pace and convenience.
- **Scalability:** Virtual visits can be easily scaled up to accommodate a large number of users simultaneously, maximizing the impact of the knowledge dissemination efforts.
- **Repeatability:** Users can revisit the virtual tour multiple times to reinforce their understanding of the agroforestry system.
- **Interactive learning:** Virtual visits can incorporate interactive elements, such as quizzes, simulations, and virtual games, to enhance engagement and promote active learning.
- **Data visualization:** Virtual visits can present complex data in an easy-to-understand format through graphs, charts, and maps, facilitating informed decision-making.
- **Environmental sustainability:** Virtual visits reduce the carbon footprint associated with travel, contributing to environmental sustainability.

Key Components of Effective Virtual Visits

To maximize the effectiveness of virtual visits, several key components should be considered:

- **High-quality visuals:** Clear and engaging visuals, including high-resolution photos, videos, and 360° panoramas, are essential for creating an immersive experience.
- **Interactive elements:** Interactive features, such as quizzes, simulations, and clickable hotspots, can enhance user engagement and promote active learning.
- **Informative content:** The content should be accurate, up-to-date, and tailored to the needs of the target audience.
- **User-friendly interface:** The virtual tour should be easy to navigate and use, even for individuals with limited technical skills.
- **Accessibility:** The virtual tour should be accessible to users with disabilities, including those with visual or hearing impairments.
- **Multilingual support:** Offering the virtual tour in multiple languages can broaden its reach and impact.
- **Expert support:** Providing access to experts who can answer questions and provide guidance can enhance the learning experience.
- **Real-time data integration:** Integrating real-time data such as weather information, soil conditions, and market prices can provide valuable insights.

Challenges and Mitigation Strategies

While virtual visits offer numerous advantages, there are also several challenges to their implementation:

- **Technological limitations:** Limited internet connectivity and access to computers or smartphones can hinder the accessibility of virtual visits, particularly in rural areas. *Mitigation:* Develop offline versions of the virtual tour that can be accessed without an internet connection. Utilize community-based learning centers with internet access.
- **Digital literacy:** Lack of digital literacy among some stakeholders can make it difficult for them to navigate and use the virtual tour. *Mitigation:* Provide training and support to help users develop the necessary digital skills. Design the virtual tour with a simple and intuitive interface.
- **Cost of development:** Developing a high-quality virtual tour can be expensive, requiring investments in hardware, software, and skilled personnel. *Mitigation:* Seek funding from government agencies, NGOs, and private sector organizations. Utilize open-source software and readily available tools.

- ☆ **Maintaining engagement:** Keeping users engaged and motivated throughout the virtual tour can be challenging. *Mitigation:* Incorporate interactive elements, gamification techniques, and personalized learning pathways. Regularly update the content and features of the virtual tour.
- ☆ **Ensuring accuracy and reliability:** It is crucial to ensure that the information presented in the virtual tour is accurate and reliable. *Mitigation:* Involve experts in the development and validation of the content. Regularly review and update the information to reflect the latest research findings and best practices.
- ☆ **Lack of sensory experience:** Virtual visits cannot fully replicate the sensory experience of being physically present at the agroforestry site. *Mitigation:* Incorporate high-quality visuals and audio, 360° imagery, and interactive simulations to create a more immersive experience.

Conclusion

Virtual visits to agroforestry models developed by ICAR-IGFRI and ICFRE hold immense potential for promoting sustainable land management practices and fostering wider adoption of agroforestry techniques. By leveraging technology to provide accessible, cost-effective, and engaging learning experiences, virtual visits can bridge the gap between research and practice, empowering farmers, researchers, policymakers, and students with the knowledge and skills they need to create a more sustainable future. Overcoming the challenges associated with their implementation through strategic planning, technological innovation, and user-centered design will be crucial for realizing the full potential of virtual visits in revolutionizing agroforestry education and extension. Future research should focus on evaluating the impact of virtual visits on knowledge adoption, behavioral change, and the overall sustainability of agroforestry systems. Furthermore, incorporating feedback from users and continuously improving the design and content of virtual visits is essential for maximizing their effectiveness and ensuring their long-term success. By embracing this innovative approach, ICAR-IGFRI and ICFRE can further strengthen their contributions to promoting sustainable agricultural practices and improving livelihoods across India and beyond.

Glossary

Agri-silviculture: A land-use system involving the conscious and deliberate co-production of agricultural and forest crops (including trees) on the same land.

Agroforestry: A collective term for land-use systems and technologies where woody perennial plants (*e.g.*, trees, shrubs, palms, bamboos) are intentionally integrated with agricultural crops and/or livestock on the same land unit, either spatially or temporally. These systems exhibit ecological and economic interactions between their components.

Agroforestry potential: The capacity within a given farming system to integrate trees or shrubs with crop or animal production, referring to their potential roles in the landscape, ditches, and arrangements of multipurpose trees (MPTs).

Alley Cropping: An intercropping system where selected species of shrubs or trees are planted in rows alongside food crops (in "alleys"). It serves as an alternative to traditional bush fallow, allowing for planned species selection and integration, often enhancing nutrient cycling and soil fertility, especially with leguminous species.

Browsing: Feeding on buds, shoots, and leaves of woody plants by livestock.

Coppice Forest: A forest regenerated by vegetative means (*e.g.*, from stumps or roots).

Coppicing: Cutting back the main stem of a tree (typically to a low height, *e.g.*, 30 cm) to induce new shoot growth.

Crop Rotation: A pattern of using land for specific crops in a defined sequence, typically to prevent soil exhaustion or to control pests and diseases.

Deciduous: A plant that sheds its leaves during a specific season, usually winter.

Dry Nursery: A nursery maintained without irrigation or other artificial watering.

Even-aged Forest: A forest composed of trees of approximately the same age, where differences typically do not exceed 25 per cent of the rotation age.

Evergreen: A perennial plant that retains its green foliage year-round, with old leaves persisting until new ones appear.

Extension Forestry: The practice of forestry in areas lacking tree growth or vegetation and situated away from conventional forest areas, aiming to increase tree cover.

Fallow: A period between crops when land is rested from deliberate cropping. Natural vegetation may be allowed to grow, or specific fallow species may be planted to provide soil cover and/or accelerate soil regeneration.

Fallow System: A cropping system in which land is rested from deliberate cultivation.

Farm Forestry: The practice of forestry on farms, including planting tree rows on field bunds or boundaries, individual trees in private agricultural land, and creating windbreaks (protective screens of trees and shrubs around a farm);

Forest: An area designated for timber and other forest products, or maintained under woody vegetation for indirect benefits such as climatic regulation or protection.

Forestry: The science and practice of growing trees, encompassing their conservation and utilization.

Frost: The chilling of air below freezing point.

Frost Hardy: A species capable of withstanding frost without being damaged;

Frost Tender: A species that is killed by frost;

Herb: A plant with a green and tender stem, usually not exceeding one meter in height. Classified as annual, biennial, or perennial based on its lifespan.

High Forest: A forest regenerated from seed.

Home Garden: An intimate association of multipurpose trees and shrubs with annual and perennial crops, and often livestock, within the compounds of individual houses, managed by family labor.

Intercropping: The practice of growing annual or short-duration crops in the interspaces of perennial crops, typically during the early years of orchards;

Light Demander: A plant that requires complete overhead light for its optimal growth and development.

Lopping: The removal of one-year-old shoots or fresh growth from a tree's entire crown, typically for livestock feed.

Mixed Cropping: Growing more than one crop species on the same piece of land at the same time;

Mixed Farming: Farming that integrates crop and animal production with functional links (*e.g.*, using manure on crops, utilizing draft animals).

Mixed Forest: A forest composed of two or more intermingled tree species within the same canopy.

Monoculture: The repeated growing of the same crop on the same land.

Mulch: Crop residues or other organic material used to cover exposed soil between growing crop plants.

Nursery: An area where plants are raised for eventual planting out.

Nursery Stock: Generally refers to plants grown in a nursery for planting out.

Phreatophyte: A deep-rooted plant that obtains a significant portion of its water from the phreatic zone (zone of saturation) or the capillary fringe above it.

Pollarding: A tree whose stem has been cut back to obtain a flush of shoots, typically beyond the reach of browsing animals;

Practices (in Agroforestry): Specific land management operations of any nature carried out on a farm or other management unit within an agroforestry context;

Pure Forest: A forest composed almost entirely of one species, usually to the extent of not less than 80 per cent.

Rotation: The time lapse between planting and the final harvesting of a plant.

Shade Bearer: A plant capable of persisting and developing under shaded conditions;

Shelterbelt: A belt of trees and/or shrubs maintained for protection from wind, sun, snow-drift, *etc.*, generally more extensive than windbreaks and sometimes covering entire regions on a planned pattern.

Shifting Cultivation: A land-use system where the cultivation site is regularly changed, with older sites reverting to forest or bush fallow;

Shrub: A woody perennial plant differing from perennial herbs by its persistence and woody stem, and less definitely from a tree by its low stature (1-6 m) and habit of branching from the base.

Silvopastoral System: The production of woody plants combined with pasture. Trees and shrubs may be grown primarily for livestock fodder, timber, fuel wood, fruit, or soil improvement.

Social Forestry: The practice of using trees and/or tree planting specifically to pursue social objectives, often improving the well-being of the poor by delivering benefits to local communities; sometimes described as "tree growing by the people, for the people."

Subsystem: A part of a larger system, with a more restricted role, content, and complexity.

Subsistence Agroforestry Systems: Land-use systems aimed at satisfying basic needs, managed by the owner or occupant and their family.

System: A group of physical components, an assemblage of objects connected or related to form and/or act as a unit (*e.g.*, an ecosystem comprises living organisms and their inseparable non-living environment).

Taungya: A method of raising forest crops in combination with an agricultural crop, where agricultural use is confined to the period before the forest canopy closes.

Tree: A large woody plant with a single, well-defined stem (bole/trunk) and a more or less definite crown, usually more than six meters in height.

Uneven-aged Forest: A forest composed of trees of markedly different ages, where individual stems vary widely in age.

Wet Nursery: A nursery maintained with irrigation or other artificial watering during dry periods.

Windbreak: A protective planting (typically 2-3 rows of trees or shrubs, spaced 0.5m-2.5m apart) around a garden, farm, or field to protect it against strong winds.

Woodlots: The practice of growing trees in very close spacing specifically for harvesting fuel wood.

References

1. Jose, S., and Shanmughavel, P. (2012). Tropical agroforestry: Tree-crop-animal interactions. Springer Science and Business Media.
2. Lin, B. B. (2007). Agroforestry management as an adaptive strategy to climate change: an example from coffee production systems. Climatic Change, 88(1), 479-494.
3. Nair, P. K. R. (1993). An introduction to agroforestry. Springer Science and Business Media.
4. Hartmann, H. T., Kester, D. E., Davies, F. T., and Geneve, R. L. (2011). *Hartmann and Kester's plant propagation: Principles and practices*. Pearson Education.
5. Macmillan, B. H. (2002). *Plant propagation: A practical guide*. Timber Press.
6. Toogood, A. (1999). *Propagation*. DK Publishing.
7. Ashraf, M., and Foolad, M. R. (2005). Pre-sowing seed treatment—a shotgun approach to improve germination, plant growth, and crop yield under stressful environments. *Advances in Agronomy*, *88*, 223-271.
8. Taylor, A. G., Allen, P. S., Bennett, M. A., Bradford, K. J., and Burris, J. S. (1998). Seed enhancements. *Horticultural Reviews*, *22*, 1-45.
9. Harman, G. E. (2000). Myths and dogmas of biocontrol. Changes in perceptions derived from research on Trichoderma harzianum T-22. *Plant Disease*, *84*(4), 377-393.
10. Farooq, M., Basra, S. M. A., Ahmad, N., Hafeez, K., and Warraich, E. A. (2005). Improving the germination and early seedling growth of hybrid

sunflower by seed priming. *International Journal of Agriculture and Biology*, 7(5), 825-828.

11. ISTA. (2024). International Rules for Seed Testing. International Seed Testing Association, Bassersdorf, Switzerland.
12. Beer, J., Ibrahim, M., and Schaller, M. (2000). *The possibilities and limitations of agroforestry in Central America*. CATIE.
13. Jose, S. (2009). Agroforestry for ecosystem services and environmental benefits: an overview. *Agroforestry Systems*, 76(1), 1-10.
14. Montagnini, F., and Nair, P. K. R. (2004). Paths to agroforestry development in tree-dominated land use systems. *Forest Ecology and Management*, 20(3), 419-423.
15. Nair, P. K. R. (1993). *An introduction to agroforestry*. Kluwer Academic Publishers.
16. Schroth, G., and Sinclair, F. L. (2003). *Trees, crops and livestock: an introduction to agroforestry*. CABI Publishing.

Index

A

Abnormal forests 15
Aesthetic enhancement 80
Afforestation method 215
Agri-horticulture 12
Agri-horti system 158
Agri-silvicultural systems 164
Agri-silviculture 13, 28
Agri-silviculture systems 28
Agro-biodiversity 55
Agroclimatic zones 247
Agroecological systems 232
Agro-ecosystem analysis 102
Agroforestry 8, 73, 233
Agroforestry management 69
Agroforestry models 163
Agroforestry planning 94
Agroforestry system 11, 25, 50, 67, 85, 91, 92, 243, 247
Agrosilviculture 28
Agro silvo pastoral systems 28
Air temperature 183
Alley cropping 12, 34, 241
Alpine scrub 17
Apiculture 49
Aqua-forestry 49
Artificial regeneration 108
Asexual propagation 114
Avenue plantations 139
Azadirachta indica 24

B

Bending 134
Bhiul/bhimal 172
Biodiversity 163
Biodiversity conservation 86
Biophysical assessments 249
Bio-priming 195
Biotic factors 118
Black cotton soils 117
Black siris 173

Bole characteristics 152
Boundary planting 165
Branch cuttings 126
Budding 210

C

Calorific value 61
Capacity building 57
Carbon sequestration 6, 74, 160
Challenges 204
Charcoal 59
Classification 26
Climate change 94
Climate change mitigation 110, 163
Climate zones 116
Coastal and deltaic alluvium 117
Cold desert 116
Communal forests 15
Context assessment 92
Contour planting 230
Coppicing 125, 134, 215
Crop combinations 37
Crop diversification 56
Crop management 68
Crop protection 80
Crop selection 185
Crop yields 40, 93, 160
Cultural practices 145, 147
Cuttings 114

D

Data collection 69, 93
Deforestation 6
Degraded forests 7
Desert soils 117
Desiccation 145
Diversified income 93
Dry nursery 139

E

Ecological aspects 45
Ecological basis 51
Ecological benefits 47, 257
Ecological restoration 110
Economic benefits 110
Economic diversification 75
Economic viability 257
Employment opportunities 6
Energy conservation 80
Energy plantation 58
Environmental parameters 181
Erosion control 40
Evaluation and adaptation 93
Evapotranspiration 40, 74
Exotic species 139
Extension forestry 7
Extractive plants 60

F

Farmer training 93
Farm forestry 7
Farm forests 14
Farming systems research 95, 102
Fertilization 68, 108, 115, 147
Field handling 146
Field preparation 198
Fire management 109, 115
Firewood 58
Fodder-fruit tree model 159
Fodder trees 48
Food 131
Food and agriculture organization 127

Food security 86
Forage alley cropping 159
Forage crops 157
Forest classification 13
Forest farming 12, 241
Forest health 109
Forest products 109
Forest regeneration 108
Forest stands 13
Forest types 16
Fuel forests 14
Fuelwood 6, 58
Fuelwood production 37

G

Genetic resource conservation 56
Gmelina arborea 25
Grafting 114, 190, 210
Grasses 39
Grevillea robusta 24
Ground operations 153

H

Handling 146
Harvesting 69, 109, 115, 153
Hedge row intercropping 35
Height and spacing 38
Homegardens 12, 48, 55, 241
Homestead agroforestry 160
Horizontal management 132
Horti-agri-pastoral system 159
Horticultural boundary plantation 159
Horticulture 157
Horti-pastoral system 158
Horti-silvicultural systems: 164
Horti silviculture 41
Horti silvopastoral 41
Hot desert 116
Humidity 40
Humid tropical 116
Hydro-priming 194

I

Indian mesquite 24
Integrated multistory systems 37
Intercropping 232
International Board for Plant Genetic Resources 128
International Council for Research in Agroforestry 128
International Tropical Timber Organization 128
International Union for Conservation of Nature 128
Irrigation 68

K

Kachnar 173
Krishishukti 3

L

Land evaluation 102
Land evaluation methodology 95
Land tenure insecurity 56
Laterite and lateritic soils 117
Layering 114, 189
Leaf cuttings 188, 190
Leucaena leucocephala 24
Light competition 46
Livelihood analysis 250
Livelihoods 86
Livestock protection 80

Local 5
Lopping 134

M

Management practices 93
Marginal lands 139
Market analysis 68, 92, 250
Market development 57
Marketing 69
Market opportunities 87
Microclimate 183
Microclimate regulation 56
Micropropagation 114
Modern planting patterns 230
Monitoring and evaluation 69, 92, 109
Monoculture reinforcement 229
Montane subtropical forests 16
Montane temperate forests 16
Mulberry 172
Mulching 203
Multipurpose trees and shrubs 36
Multipurpose tree species 23, 130
Mycorrhizal associations 46

N

National and regional institutions 129
Natural and artificial regeneration 237
Natural regeneration 108, 238
Nature's fertilizer factory 46
Neem 24
Nettle khirk 174
Niterative process 97
Nitrogen fixation 46
Nursery bed 137, 201
Nursery management 211
Nursery stock 145
Nursery stock preparation 146
Nutrient competition 46
Nutrient content 183
Nutrient cycling 119
Nutrient management 69
Nutrient recycling 45
Nutri-priming 195
Nutritional and water requirements 153

P

Panchayat forests 15
Participatory rural appraisal 92, 250
Permanent nursery 140
Pest and disease management 69, 109, 115, 147
Pest and disease outbreaks 204
Phenology 153
Planning and design 67
Plant establishment 39
Planting and crop management 198
Planting depth 146
Planting pattern 227
Planting techniques 146
Planting time 146
Plant propagation 187
Policy and institutional support 87
Pollarding 134, 215, 221
Post-planting care 218
Practical application 256
Productive functions 17
Productivity 5
Propagation and management 113
Propagation techniques 207
Prosopis juliflora 24
Protected forests 15

Protective and ameliorative functions 18
Protective forests 14
Protein banks 47
Pruning 68, 108, 115, 133, 215
Pruning and thinning 147

R

Rainwater harvesting 74
Recreational and educational functions 18
Recreational forests 14
Recreation forestry 7
Red soils 117
Regeneration method 14
Relative humidity 183
Reserved forests 14
Resins 60
Resource inventory 92
Rhizomes 126
Root cuttings 189, 190
Root suckers 126

S

Saline-alkali soils 117
Seed collection 113
Seedlings 171
Seed material 197
Seed production 138
Seed propagation 113, 188, 190
Seed treatment 193
Sequester carbon 47
Sesbania 25
Sexual propagation 113
Shaping growth 219
Shelterbelt 79
Shelterbelts 38
Shifting cultivation 28, 29, 56
Shisam 175
Shrubs 39
Silver oak 24
Silviagriculture 28
Silviculture 11, 107, 151
Silvi-horti system 159
Silvi-pastoral systems 47, 164
Silvo-pastoral systems 28, 241
Silvopasture 11
Site assessment 244
Site preparation 114
Site selection 202
Slow-growing species 138
Small trees 39
Snowdrift control 80
Soapnut/ritha 174
Social forestry 8
Socio-economic aspects 85
Socio-economic assessments 249
Socioeconomic basis 51
Soil and water conservation 73
Soil conservation hedges 40
Soil degradation 204
Soil erosion 46, 80
Soil health 6, 47, 56, 160, 163
Soil improvement 217
Soil management 68, 185
Soil moisture 40
Soil moisture content 183
Soil nutrients 118
Soil organic matter 183
Soil preparation 38, 203
Soil texture 182
Spatial arrangement 93
Species composition 14

Species composition management 109
Species diversification 185
Species selection 93, 208
Square or rectangular pattern 228
Stand dynamics 108
Stem cuttings 188, 190
Strip cropping 231
Stump sprouting 126
Sub-alpine forests 16
Sub-system 25
Sub-tropical region 59
Subtropical semi-arid 116
Subtropical semi-humid 116
Sustainable future 75
Sustainable resource management 110
System design 93, 132

T

Taungya system 12, 31, 56
Temperate climate 59
Temperate semi-arid 116
Temperature moderation 40
Temporal management 133
Temporary nursery 139
Tending 153
Thinning 68, 108, 114, 133
Topography 216
Traditional agroforestry 55
Traditional knowledge 56
Traditional planting patterns 228
Training 134, 215
Transplanting 126
Transportation 146
Treatment methods 198
Tree-crop dynamics 45
Tree-crop interaction 45
Tree growth measurement 177
Tree management 185
Trees 39, 207
Tree selection 178
Tree species 59, 113, 171
Tropical dry region 58
Tropical forests 16
Tropical humid region 59
Tropical rainforests 248
Tropical semi-arid 116
Tropical sub-humid 117

V

Value chains 87
Vegetable oils 60
Vegetative propagation 114
Vertical management 133
Village forests 15

W

Wasteland 6
Water competition 46
Watering 147
Water management 69, 115, 131
Water quality improvement 74
Water scarcity 204
Weed competition 204
Weed control 68, 114, 147
Weeding 109
Wet nursery 139
Wildlife habitat 80
Windbreak 79
Windbreaks 40
Wind speed 183
Wind velocity 40
Women 86
Woody hedges 49